Technoscience and Environmental Justice

Urban and Industrial Environments

Series editor: Robert Gottlieb, Henry R. Luce Professor of Urban and Environmental Policy, Occidental College

For a complete list of books published in this series, please see the back of the book.

Technoscience and Environmental Justice
Expert Cultures in a Grassroots Movement

edited by Gwen Ottinger and Benjamin Cohen

afterword by Kim Fortun

The MIT Press
Cambridge, Massachusetts
London, England

For information about special quantity discounts, please e-mail special_sales@mitpress.mit.edu

This book was set in Sabon by Toppan Best-set Premedia Limited. Printed and bound in the United States of America.

Library of Congress Cataloging-in-Publication Data

Technoscience and environmental justice : expert cultures in a grassroots movement / edited by Gwen Ottinger and Benjamin Cohen.
    p.   cm. — (Urban and industrial environments)
    Includes bibliographical references and index.
    ISBN 978-0-262-01579-0 (hardcover : alk. paper) — ISBN 978-0-262-51618-1
(pbk. : alk. paper) 1. Technology—Environmental aspects—Case studies. 2. Science–
Environmental aspects—Case Studies. 3. Industries—Environmental aspects—Case
Studies. 4. Environmental justice—Case studies. 5. Environmental health—Case studies.
6. Environmental policy—Citizen participation—Case studies. 7. Environmental policy—
Decision making—Case studies. I. Ottinger, Gwen. II. Cohen, Benjamin R.
    TD194.T43   2011
    363.7—dc22
                                                                      2010048351

10   9   8   7   6   5   4   3   2   1

Contents

Acknowledgments

While we were making the final revisions to this book, one of us (Ottinger) was contacted by Peter Wood, a scientist from the California EPA's Department of Toxic Substance Control. Describing a community suffering from headaches and nosebleeds that they attributed to a methane leak from a nearby facility—a community whose concerns had been dismissed by the local Air Quality Management District—Wood explained that he was looking for resources that would help him solve the community's problem. Indeed, as it turned out, Wood had coauthored a concept paper laying out a new model for agencies' work with environmental justice communities, a model that centered on community problem solving.

Wood's call was evidence of the timeliness and importance of the chapters collected in this book. Every day scientists push to create and recreate strategies for doing their jobs in ways that further environmental justice; they do so in conversation with community groups, often talking across cultural divides. Moreover, Wood was a reminder of the gratitude we owe scientists and engineers engaged in environmental justice advocacy. Over the course of our respective research projects and our teaching at the University of Virginia, we have gotten to know a number of exceptional scientists and engineers committed to environmental justice, including Julia May, Azibuike Akaba, Wilma Subra, and Don Gamiles, to name a few. We thank them for inspiring this book, as well as for their service to communities.

Of course, this book would not have been possible without the support of generous colleagues. We are grateful to Bernie Carlson for encouraging the project and helping us to find the resources to carry it out, and we thank Kim Fortun for her thoughtful mentorship at crucial moments in the process. We are also terribly fortunate to have had a group of contributors who, in addition to being intellectually engaged, found myriad ways to offer their encouragement, support, and friendship over the years during which the book took shape.

Finally, we wish to thank the Darden School of Business, the Department of Science, Technology, and Society, the Department of Environmental Sciences, and the Department of Anthropology, all at the University of Virginia, for funding a two-day workshop for contributors. Ottinger is also grateful to the Environmental History and Policy Program at the Chemical Heritage Foundation for a postdoctoral fellowship that gave her time to complete this project.

Introduction: Environmental Justice and the Transformation of Science and Engineering

Benjamin Cohen and Gwen Ottinger

For nearly thirty years, the environmental justice (EJ) movement has been engaged in what Cole and Foster (2001) describe as "transformative politics." In the course of agitating to correct inequities in the distribution of environmental hazards, the movement has transformed the victims of environmental injustices, turning formerly quiescent minority and low-income neighborhoods into organized, politically engaged communities, and residents once intimidated by powerful corporations and state institutions into outspoken, politically savvy advocates for their communities. Environmental justice activism has also played a role in changing environmental policy, as U.S. agencies have been forced to consider the health and environmental consequences of their decisions for communities of color and low-income communities (see Cole and Foster 2001, 26; Toffolon-Weiss and Roberts 2005; Gordon and Harley 2005; NRC, 2009; NAS, 2009). In addition, the environmental justice movement has transformed the shape of environmental activism, calling environmentalists' attention to the places where we "live, work, and play," as leading EJ advocate Robert Bullard put it, and pioneering a networked, grassroots structure that arguably enables the movement to remain community-driven (Schweizer 1999; Schlosberg 1999; Agyeman 2005).

Expanding the notion of environmental justice as transformative politics, this book contends that the environmental justice movement has also been transforming science and engineering. The book's chapters present case studies of technical experts' encounters with environmental justice activists and issues, inquiring into the transformative potential of these interactions. The contributors ask: To what extent has working with EJ activists enabled scientists and engineers to forge new scientific practices and identities? What has constrained their ability to do so? They also ask parallel questions of technical professionals' activities in mainstream scientific institutions, including places such as regulatory agencies and universities, where commitments to environmental justice have been (partially) institutionalized in a variety of ways: How, and in what ways, have the practices of science and the identities of scientists within scientific institutions changed as a result of pressure from the EJ movement? What are the structures that have stood in the way of more fundamental change?

In addressing these questions, we draw on an understanding of science and engineering as multivalent with respect to environmental justice. Studies of environmental justice activism have highlighted the central roles that technical practices and technical experts often play, not only in creating environmental injustices, but in confronting and correcting them as well. The literature, that is, documents the heterogeneity of environmental justice activists' engagements with experts and their knowledge. It demonstrates that the activities of technical professionals may contribute to the structural inequities that put communities of color and low-income communities in harm's way while at the same time offering resources in those communities' pursuit of justice. Nor are these possibilities mutually exclusive: even scientists seeking to aid community groups may inadvertently reproduce the structures of injustice that they seek to counter (Cable, Mix, and Hastings 2005; Hoffman, chapter 2, this volume).

Besides being multivalent and heterogeneous in the context of environmental justice, science and engineering are dynamic by nature and therefore transformable. This book contends that the dynamism of technical practices has gone largely unacknowledged by earlier EJ research. Researchers have highlighted the culturally contingent nature of science and the values, biases, and systemic blind spots inherent in it through, for example, studies that contrast community understandings and representations of local environmental conditions with "scientific" understandings and representations (Wynne 1996b; Bryant 1995; Head 1995; Tesh 2000). However, scientific knowledge is largely regarded as relatively stable in this work, its shortcomings predictable and enduring, rather than as an ongoing cultural creation, made and remade through the daily practices of scientists and engineers. Researchers have tended to represent alternative forms of science, if found at all, as challenges to—rather than extensions or refashionings of—the official knowledge produced in mainstream scientific institutions (e.g., Brown 1992; Corburn 2005). Those institutions along with the scientists and engineers who populate them are even less likely to be characterized as dynamic and changeable; for the most part, EJ studies present technical practitioners as static, external agents who "parachute in" to predetermined scenarios (see Irwin, Dale, and Smith 1996).

Perhaps because they are viewed as static, scientific knowledge, institutions, and experts have largely been excluded from accounts of environmental justice's transformative nature. Works that document ordinary citizens' journeys from quiet homemakers to outspoken activists (e.g., Cole and Foster 2001; Pardo 1998) lack counterparts describing allied scientists' journeys from bench researchers to experts willing to advocate on communities' behalf (though Allen 1998 is a partial exception). Similarly, research that asks how the environmental concerns of poor and minority communities have changed the shape of the "mainstream" environmental movement (Schlosberg 1999; Brulle and Essoka 2005) has not been extended to consider how government acknowledgment of environmental justice concerns has changed the shape of environmental regulatory institutions and the science practiced therein.

In this book, we extend researchers' and practitioners' understanding of environmental justice as a transformative movement by showing how environmental justice activism has created opportunities for changing technical practices—and, in a few cases, has even made possible significant transformations. Drawing on insights from science and technology studies (STS), we offer a theory of how science and engineering can change—namely through technical professionals' responses to routine "ruptures" in their practice. The size of the ruptures, we suggest, determines the extent of science's dynamism—that is, the degree to which institutional contexts allow scientists room to maneuver limits their ability to bring about fundamental change.

Our theory of rupture and room to maneuver provides a framework for understanding the transformations—achieved and hoped-for, thorough and incomplete—explored in the case studies that comprise the book. Each chapter documents a rupture, an opportunity for change, created through the interactions of scientists and engineers with environmental justice activists and issues. Each also points to the ways that the larger structures that constitute the contexts for scientists and engineers' work either help them to take advantage of those ruptures or, more often, constrain their room to maneuver and stand in the way of fundamental transformation. By rendering technical practices as potentially malleable, this book suggests how they might be remade in the service of environmental justice. Beyond our theory of rupture and constraint, the individual chapters suggest strategies that might be employed in order to bring about change, and they detail obstacles in order that activists and practitioners may look for a way around them.

In the introductory remarks that follow, we first discuss the variety of ways in which science and technology have been engaged by the environmental justice movement. We then elaborate on how work documenting the dynamic, contingent character of science and technology allows us to identify and analyze ways that engagement in the environmental justice movement has opened up spaces for transformation of the scientific enterprise. In the subsequent section, we discuss the two classes of transformation that divide the book into two parts—those initiated by scientists and engineers and those provoked by pressure from EJ advocates—before we make the ideas of rupture and constraint, opportunity and obstacle, concrete through a discussion of how the twinned concepts appear as driving themes in the chapters, even if not invoked explicitly by contributors. In the end, we argue for the value of understanding technical practices as multivalent and dynamic, showing that environmental injustice is an important source of ruptures in technical practice and thus a powerful force for the transformation of science.

Environmental Justice's Multivalent Engagement with Expertise

Research and advocacy around environmental justice issues have made clear that environmental racism and injustice are structural problems (Bullard 2000; Cole and Foster 2001). Focused on instances where community groups have successfully mobilized against these structures, researchers and advocates have also stressed that environmental injustices can be overcome (Pellow and Brulle 2005; Lerner 2005). In activists' battles to dismantle the structures of environmental injustice, we suggest, technical expertise appears on both sides. Scientific practices, technological infrastructures, and the authority of technical experts all help produce and reinforce the structures that create environmental injustices. At the same time, science and expertise have also been powerful weapons in struggles to overcome environmental injustices.

Industrialization and its associated sociotechnical infrastructures have both produced the environmental hazards faced by communities and shaped practices for generating and acting on scientific knowledge. In the post–World War II era of Big Science, cycles of industrial production, consumption, and disposal grew on an unprecedented scale. The increase in production-consumption cycles throughout the twentieth century has been the source for the environmental hazards that communities, including those instrumental in the rise of the EJ movement, have organized to confront. The polychlorinated biphenyl (PCB)–contaminated soil that African-American residents of Warren County, North Carolina, fought to keep out of their community in 1982, for example, originated as a by-product of industrial processes to make components for electrical systems (such as insulators and heating fluids) and plasticizers for a generation of new polymeric materials (Bullard 2000). The leaking drums of chemical wastes that a white community in Love Canal, New York, fought to have cleaned up in the late 1970s had their genesis thirty years prior with the closing of a chemical production facility (Blum 2008). The construction and expansion of petrochemical facilities in the latter half of the century had ramifications for communities, as in the case of Norco, Louisiana, where a company's decision to build a new facility in the 1960s sandwiched a historic African-American community between a chemical plant and an oil refinery (Lerner 2005).

Such cases comprise a central plank in an environmental justice movement, where minority and low-income communities have fought new sources of pollution that added to their disproportionate burdens, insisted on cleaner and safer processes at existing production and processing facilities, demanded that environmental laws be fully enforced in their communities, and defended their right to participate in decisions that would affect the health of their families and neighborhoods. When viewed together, the cases show how historical patterns of residential segregation have combined with local zoning ordinances to make hazardous facilities more likely to be sited near communities of color or how residency patterns encourage particular land-use policies by local governments. They illustrate practices used to silence low-income, minority, and immigrant communities and privilege corporations and wealthier citizens in policymaking processes (Cole and Foster 2001; Shrader-Frechette 2002). They speak, that is, to the structural issues at play in environmentally unjust scenarios.

The growth of twentieth-century industry was intertwined with another structural contributor to environmental injustice: the rise of an expert class whose claims to authority were, and are, based in specialized, technical training. While "the scientist" and "the engineer" both developed professional identities before the twentieth century, they acquired new status in a modern industrial system that relies on science and technology to generate wealth. The authority of technical experts has, paradoxically, been bolstered by the environmental hazards associated with the industrial system: unprecedented environmental problems have created a situation—characterized by Ulrich Beck (1992) as a "risk society"—in which only science and technology can provide solutions to the problems created by science and technology. In the risk society, as Pellow and Brulle (2005) point out, technical experts are deeply implicated in creating and maintaining environmental injustices. Experts' calculations, political and economic as well as technical, inform the siting and enforcement decisions that concentrate environmental risks in low-income communities and communities of color. Called to help solve environmental problems, scientists and engineers are likely to favor research aligned with the goals of the laboratory or the agency or the corporation, not the community beyond the fenceline (Hess 2007, 2009). The increasing influence of neoliberal agendas, where market-based solutions define the form of response to environmental problems, further embeds science and engineering within structures of capitalist production (Liévanos, London, and Sze, chapter 8, this volume). They also value certain kinds of evidence and discount others; the "local knowledge" or "citizen science" that community groups could bring to environmental problems, for example, is frequently disregarded or deemed irrelevant in the context of mainstream scientific practice (Ottinger 2010).

In concrete terms, experts' centrality to the modern industrial system has been directly consequential for environmental justice struggles. Communities seeking to demonstrate that their health is (or would be) harmed by nearby hazards often find little help in accepted modes of scientific practice. Inquiry into the potential health effects of chemical exposures, where it occurs at all, is constructed in a way that makes it extremely difficult to show that chemicals cause illness in humans under the conditions experienced by communities engaged in environmental justice struggles. Epidemiological studies, for instance, routinely miss or even deliberately obscure the effects of environmental hazards on community health (Lewis, Keating, and Russell 1992; Head 1995; Tesh 2000). Technical experts' participation in public hearings and other decision-making processes can limit citizens' ability to have their voices heard (Gauna 1998). Nor are the cases only the result of status quo passivity, since research by government agencies in a number of influential cases was designed specifically to obscure any connections between illness and pollution (Allen 2003; Lewis, Keating, and Russell 1992).

The dominant place of experts and expert knowledge also constrains communities' ability to see local environmental hazards addressed. Policymakers will rarely take action to mitigate the effects of pollution on a community until they have proven that pollution *causes* residents' health issues—an almost insurmountable hurdle given scientists' standards for proof (Bryant 1995). Simultaneously, with technical information about facilities and their environmental risks given privileged status in siting decisions and other public policy processes, residents frequently find their ability to participate fairly in those processes circumscribed. Community members' lack of familiarity with highly technical terminology can limit their ability to challenge the bases on which policy decisions are made, since their contributions to public hearings and other deliberations are likely to be devalued for not being "scientific." Compounding the barriers faced by communities is a regulatory regime that draws from the technical expert class to set and define standards based on scientific norms produced in labs and outside affected communities.

Intertwined with modern industry, science, technology, and expertise are thus integral to the social and political structures that produce environmental injustices. Accordingly, EJ advocates have in many cases taken expert knowledge as a target. In particular, they have challenged claims that communities are not harmed by industrial pollution. Arguing that environmental health science routinely fails to incorporate relevant "local knowledge," EJ advocates emphasize that those situated understandings of local environmental conditions and behavioral patterns are likely to influence community exposures to toxins (DiChiro 1998; Sze and London 2008). They have also shown that scientists' ignorance of or faulty assumptions about local conditions can result in significant distortions in scientific studies and risk assessments (Harris and Harper 1997; Corburn 2005; see also Powell and Powell, chapter 6, as well as Johnson and Ranco, chapter 7, this volume).

Yet, as we noted above, EJ activists have also used science as a resource in mobilizing against the structures of injustice. Communities have enlisted the support of experts sympathetic to residents' claims about the health effects of environmental hazards; some have even worked with experts to produce alternative studies that capture residents' local understandings of environmental and health problems (Brown and Mikkelson 1990; Brown 1992; Singleton and Legator 1997). Expert activists have, in conjunction with community members, developed other new methods for representing patterns of illness in polluted communities, including the mapping techniques that Barbara Allen (2000, 2003) describes. In other cases, community groups have used environmental monitoring technologies to produce scientific data about their exposures to chemicals as a way to hold industrial facilities and environmental regulatory agencies accountable for chemical pollution (O'Rourke and Macey 2003; Overdevest and Mayer 2008; Ottinger 2010). These efforts draw on the authority of science and scientists to further communities' environmental justice goals. Yet they do not merely reproduce established scientific practices. Rather, in mobilizing science for their own ends, grassroots groups have been creating alternative methods for knowing about and representing the health effects of pollution. The epidemiological, monitoring, and mapping methods that EJ advocates have pioneered provide technical information that represents local knowledge and fills gaps in existing science.

We can see environmental justice activists' effort to turn science to their own ends, creating new knowledge and practices in the process, in one of two ways. Their efforts can be regarded as a unique endeavor distinct from and opposed to conventional forms of science and expertise. But they can also be seen as an extension of established scientific practices that aims to alter the shape and direction of those practices. In the first view, the alternative forms of science created in the service of environmental justice are combative; in the second, they are transformative. We see communities' engagements with science, technology, and technical experts as having the potential to transform scientific practices and methods, a view that raises new questions for the study of relations between science and environmental justice. What are the processes through which transformation takes place? Are there particular points of leverage that can be used to facilitate change, or stumbling blocks that stand in the way of efforts to alter technical practices? And how can EJ advocates orient their efforts to promote the transformation of science so that it becomes less often an obstacle to environmental justice and more often a resource?

Constructing and Reconstructing Science

Asking how science can be—and has been—transformed by environmental justice advocacy depends, first, on understanding science as active and evolving. Too often in the cases noted above, scientific knowledge has been regarded as relatively stable, its shortcomings predictable and enduring. Technical practitioners, likewise, tend to be represented as static, external agents parachuting in to communities to apply pre-established skill sets. In contrast, we see science and engineering as flexible, contingent, and continuously under revision. In these dynamic enterprises, transformations grow out of routine ruptures in everyday technical practices, where scientists and engineers have room to make new choices about how to do their work.

Viewing science as dynamic and subject to constant revision, rather than fixed and stable, follows from the insight that science is a culturally situated set of practices.[1] Developed in the decades since Thomas Kuhn's ([1962] 1996) postpositivist history and philosophy of science, this understanding of science contrasts with older, though more publicly familiar, positivist conceptions of science. That earlier work conceived of an asocial, rational science devoted to uncovering timeless and place-less facts about the natural world (see Turner 2008). Since Kuhn, numer-ous studies of the actual practices of scientific knowledge production have come to find that science is a more complicated, more culturally embedded, and more dynamic social system than positivist portrayals can account for. Notably, this research has argued that scientific "facts" do not enter the social world fully formed; rather, they are produced in the course of working out other kinds of political and social arguments (see Sismondo 2004, 2008; Yearley 2005). Nor can the development of technology be understood as the straightforward application of scientific principles to preestablished problems; technology too is shaped by a variety of social and political negotiations that simultaneously define an object's form, its meaning, and the societal problems to which it is a solution (Winner 1986; Bijker, Hughes, and Pinch 1987; MacKenzie and Wajcman 1999; Oudshoorn and Pinch 2003).

Science and technology are actively constructed through the efforts of a variety of actors embedded in specific relations of power and systems of cultural meaning. Without suggesting that the situatedness and historical contingency of science and technology are a priori good or bad, the finding has directed scholarly attention to the actual processes through which facts and artifacts are made (see, for example, Daston and Galison 2007 on objectivity; Latour and Woolgar [1979] 1986, Knorr Cetina 1981, and Lynch 1985 on microlevel laboratory studies; and Frickel and Moore 2006, Hess 2007, and Hess 2009 on meso- and macrolevel institutions and networks). Scholars' interest in understanding science as a culturally situated production also extends to exploring the very processes through which scientific practitioners are constructed as "experts." Research on boundary drawing and the creation of scientific personae, specifically, document how scientists claim authority over particular domains of knowledge (Gieryn 1999; Daston and Sibum 2003; Carson 2003; Browne 2003; Thorpe and Shapin 2000).

Understanding science, technology, and expertise as actively produced rather than pregiven indicates the dynamism of scientific and engineering enterprises. Science is not only made, it is constantly in the process of being remade in response to shifts in cultural terrain. The active character of technical practice suggests the possibility for scientists to deliberately refashion their practices, institutions, and identities to bring about changes in the nature of scientific research or the bases for expert authority. Scholars have traced such developments in the last few decades with examples ranging from the fields of genetic toxicology (Frickel 2005) and conservation biology (Galusky 2000) to individual "politicoscientists" in the post–World War II era (Egan 2007; Moore 2008). Even more broadly, Shapin (2008) has shown how the very idea and moral constitution of science and "the scientist" have evolved into a vocation in the past century. Nor is making and remaking science the province of practitioners alone. Epstein (1996) and Hess (2007), for instance, have chronicled the effects that social movement groups have had on the shape of scientific research and technological development; Nowotny, Scott, and Gibbons (2001) argue in their explication of "Mode 2" science that research agendas are now set beyond the confines of disciplinary homes, even if not by the social movement actors Epstein and Hess refer to.

Although science and technology are always in the process of reconstruction, we argue that significant transformation has the greatest opportunity to occur at moments where there are disruptions in everyday technical practices. As a social practice, science is constructed through the everyday choices and activities of practitioners as they go about their work (Bourdieu 1977; De Certeau 1984). Those activities are always aided, constrained, and given meaning by the broader world in which practitioners are situated—in fact, the reconstructions are mostly shaped in subtle ways within power structures of dominant institutions—yet these situations always come with indeterminacy, and thus maneuverability, in their choices. The bounded but indeterminate nature of practice creates rifts, ruptures, or "spaces between" that, like an earthquake breaking apart the land or a knife slicing a loaf of rising bread, produce more area than existed before (Serres 1983; Deleuze and Guattari 1987; Traweek 2000). These are sites of transformation and, ultimately, the source of the flexibility and dynamism that STS scholars have shown to be characteristic of science.

Environmental justice activity is a powerful source of ruptures and "spaces between" in technical practice. Where scientists and engineers respond to the demands of stakeholders, affected users, and collaborators from the environmental justice movement, the broader world that aids and constrains their work expands, spaces open up, and the choices available to them multiply. By heightening the indeterminacy of technical practices, experts' involvement with EJ expands the opportunities for transformation. The transformative opportunities introduced by environmental justice are not unbounded, however. Technical practices remain constrained by the institutions and power relations within which scientists and engineers operate, circumscribing the possibilities for transformation.

These opposing dynamics—of opportunity and obstacle—are made concrete by the case studies collected here. In chronicling various kinds of interactions between technical practitioners and environmental justice advocates, they show how EJ engagement, voluntary or involuntary, generates ruptures and broadens experts' room to maneuver. Simultaneously, they highlight systemic and structural obstacles experts face in pursuing transformative opportunities. Individually and together, they suggest ways that environmental justice advocates' engagements with science can be made more effective in changing technical practice.

Rupturing Practice from within and without: Opportunities and Obstacles

The studies assembled in this book examine ruptures in technical practice created by environmental justice activity. Focusing on a moment or site where technical practitioners meet activists, issues, and informational needs associated with the EJ movement, the chapters explore how these interactions create opportunities for transformation, analyze the degree to which change is realized, and theorize the factors that foster or limit change.

The diversity of the collected case studies points to two sources of rupture. The chapters in part I of the book discuss opportunities for transformation that occur as a result of experts' deliberate efforts to support the broader movement for environmental justice through their technical practices. In contrast, chapters in part II highlight how science and engineering practice within mainstream institutions has been disrupted by outside pressures on the institutions to incorporate environmental justice sensibilities into their work. In the first half, the examples illustrate cases of invited change; in the second half, they show cases where changes were not invited.

The authors show that the transformative potential of both kinds of rupture stands in tension with constraints created by the institutional contexts and cultural milieus in which technical practitioners operate. While the obstacles to change within established institutions such as regulatory agencies and undergraduate training programs are all too predictable, chapters in part II nonetheless describe indications of important shifts in technical practices and identities that have occurred as a result of EJ activism. Conversely, the case studies in part I find that even where scientists and engineers actively seek change, their ability to transform their practices may be constrained by the nature of their scientific networks, by accepted standards of practice, and by their own understanding of their role as experts.

The transformative interventions of scientists and engineers showcased in part I include pioneering research in environmental health science, the focus of Scott Frickel's chapter (chapter 1) exploring how scientists come to be expert activists in the EJ movement; a nonprofit organization's scientists' efforts, analyzed by Karen Hoffman in chapter 2, to share their expertise and the political influence that goes with it with community members without formal technical training; the development of websites to make information about the releases of toxic chemicals into communities widely available, critically examined by Jason Delborne and Wyatt Galusky in chapter 3; a collaborative project among researchers (including the lead author of chapter 4, Rachel Morello-Frosch), environmental justice groups, and community partners to provide community members with meaningful information about the levels of toxic chemicals in their bodies; and a project by Sri Lankan engineers to promote renewable energy development in the country's rural communities, analyzed by Dean Nieusma in chapter 5.

The heterogeneity of those cases emphasizes the range of settings in which scientists, engineers, and other experts find and create opportunities for change. Cases of toxicologists, epidemiologists, and other scientists cooperating with community groups to develop new methods for investigating environmental health problems—the kinds of cases invoked by Frickel and Morello-Frosch et al.—have become relatively familiar (e.g., Allen 2000; Brown 1992, 1997; Corburn 2005). But these chapters additionally show energy engineers (Nieusma) and web developers (Delborne and Galusky) attempting to deploy their particular skills in ways that support the goals of environmental justice. Moreover, they show experts working to reshape multiple aspects of technical practice—not only processes of knowledge production but also methods of disseminating technical information (Morello-Frosch et al.; Delborne and Galusky) and relationships between experts, community members, and policymakers (Hoffman; Nieusma).

Simultaneously, part I's chapters detail systemic challenges faced by scientists and engineers wishing to change various aspects of their practice. Some of the work highlights experts' ongoing connections to pre-existing professional networks and identities as particularly significant in limiting the success of transformative efforts. New approaches to reporting research results back to community groups, for example, must contend with entrenched understandings of ethical practice, as Morello-Frosch and colleagues show. Likewise, scientists and engineers' sincere efforts to empower community members by transferring technical information and authority are, in the cases developed by Delborne and Galusky as well as Hoffman, constrained by their own preconceptions of the respective roles and abilities of technical experts and "laypeople" in achieving environmental justice ends. The chapters by Frickel and Nieusma underscore the importance of professional networks by exploring the particular conditions under which scientists and engineers can use their networks to help them be effective environmental justice advocates. Identifying and characterizing such challenges provide a necessary precondition for working toward future transformations. If the goal is to discuss ruptures and flexible joints in technical practice, then it is also necessary to mark out locations of particular intractability so that, in the future, practitioners can feel their way around the situation like a demolition crew looking for weak points in a building, seeking new chances to exploit break points yet unfound.

The second half of the book focuses on the transformative potential of *un*invited ruptures produced by environmental justice activity, showing in a range of cases how scientists and engineers are adapting to EJ advocates' push for new kinds of technical practice. Three of the chapters focus on the regulatory arena. In chapter 6, Maria Powell and Jim Powell document their EJ organization's efforts to change regulators' approach to evaluating and providing information about the health risks of subsistence fishing by minority communities in the polluted lakes of Madison, Wisconsin; in chapter 7, Jaclyn R. Johnson and Darren J. Ranco discuss the ways that the EPA has engaged Native American communities and scientists in a joint effort to create culturally appropriate methods of risk assessment; and in chapter 8, Raoul S. Liévanos, Jonathan K. London, and Julie Sze analyze how the California Department of Pesticide

Regulation accommodated the concerns of environmental justice activists, including calls for attention to the cumulative impact of assorted environmental hazards and an ethic of precaution, into its study of the effects of pesticides on the Latino community of Lindsay, California. The section's fourth chapter (chapter 9) considers ruptures in the training and socializing of technical professionals. There, Gwen Ottinger discusses undergraduate engineering students' responses to term projects that assigned them a role as expert activists for communities engaged in environmental justice struggles.

Like those in the first half of the book, the chapters in part II also point out obstacles to change in mainstream technical practices. They detail how regulatory structures obscure differences of race and culture (Powell and Powell; Johnson and Ranco) and resist fundamental changes to established paradigms of risk assessment (Johnson and Ranco; Liévanos, London, and Sze). They highlight how differences in power between regulatory agencies and community groups are an important obstacle to fundamental change (Johnson and Ranco), as is the robust "expert" identity that scientists and engineers are encouraged to develop as part of their specialized training (Ottinger).

Despite the barriers to change, the chapters in part II also document not only opportunities for, but actual moments of, transformation in technical practice. Describing instances where the practices of regulatory scientists have incorporated the principles of environmental justice (Johnson and Ranco; Liévanos, London, and Sze), these chapters offer evidence that institutional commitments to EJ won through social movement activism have affected the way that scientific research is both conducted and used in policy processes. The chapters also draw attention to technical practitioners' growing awareness of environmental justice issues, highlighting personal transformations as an important part of the process of transforming the practices and institutions of science (Ottinger; Powell and Powell). Even where the transformations in part II are uneven and authors critique the ways EJ advocates fall short of the fundamental changes they imagine, the chapters offer hope and optimism about the dynamism of all technical practices—not just those explicitly allied with EJ—and the transformative potential of ruptures created by environmental justice activism. In her afterword, Kim Fortun echoes this optimism, adopting the metaphor of "faultlines" to amplify the book's theme of ruptures and breaks. It is where faultlines appear, Fortun observes, that advocates have the chance to produce new thinking, new policies, and new practices.

Science and the Transformative Politics of Environmental Justice

Through their activism and scholarship, environmental justice advocates have shown how science-as-usual tends to stand in the way of EJ goals. Their work makes clear that achieving environmental justice requires transforming scientific and technological practices. This book aims to contribute to the proactive reconstruction of science—a project shared by STS scholars concerned with making science more democratic[2]—by understanding the processes through which transformation has and could occur, along with the obstacles that stand in the way of change. In the process of providing grounded analyses of technical practitioners' engagements with environmental justice, the authors suggest hypotheses for EJ scholars to test in future research and lessons for practitioners in similar situations wishing to maximize the transformative effects of their activities.

In addition to the insights of the individual chapters, the book as a whole offers three key insights of value to environmental justice advocates. The first and most straightforward is that technical practitioners' engagement with environmental justice issues—and EJ activists' engagement with science, technology, and experts—do produce ruptures in technical practice. These, in turn, make space available for culturally authorized experts in nonprofits, universities, and regulatory agencies alike to build new practices and to alter the scientific enterprise more generally. In the case of the biomonitoring studies described by Morello-Frosch et al., collaborating with environmental justice groups forced scientists to question the ethical frameworks in place for reporting data back to the individuals from whom it was collected. Collaboration between scientists and activists thus produced a new ethical framework for reporting results. In the collaboration analyzed by Hoffman, scientists from a nonprofit organization allied with community groups reconsidered how they should play their role as "experts" in public hearings in order to give community members a potentially greater voice in the proceedings. For engineering students in Ottinger's chapter, involvement with EJ activists suggested the need to develop new approaches to analyzing data—approaches that would help represent the experiences of community members in quantitative terms—as well as the need to include activists in problem solving in order to incorporate those experiences. Among other innovations documented in the following chapters, these new approaches to scientific ethics, to data analysis, and to participation in knowledge and policymaking processes were made possible—and indeed necessary—through interactions with activists that introduce new demands and new resources to technical practice.

The finding that technical practitioners' engagements with environmental justice can be transformative has clear practical implications. Chief among these is that interested actors can and should create more areas of contact between scientists, engineers, and the environmental justice movement—and, in so doing, actively nurture networks of expert activists—so as to expand the spaces available for rethinking technical practices. Such contact needs to be encouraged, fostered, and instigated wherever possible, be that in classrooms, public meetings, instances of regulatory reform, or community organizing forums. Certainly collaborations of this nature can be difficult, but even where the transformative opportunities of such interactions are not fully met, they are worth pursuing for the room to maneuver that they create. It is incumbent on practitioners of today to foster such spaces where future advocates can operate.

The collection's second conclusion is that transformations of power relations are integral to scientific transformations. The authors' analyses of obstacles to change make clear the importance of power relations, particularly the constraints that limit technical practitioners' room to reconfigure their activities. In Johnson and Ranco's account in chapter 7, for example, the new and culturally relevant concept of "health" offered by EPA's Tribal Science Council ultimately had little impact on regulators' practices of environmental risk assessment because the agency was unwilling to share its decision-making power. Similarly, in the case described by Hoffman in chapter 2, the status afforded to authoritative experts in public hearings made it difficult for scientists who were also EJ advocates to facilitate the participation of nonexpert groups. In these and other cases, efforts to create new ways to generate, analyze, or make available scientific data fell short, in the authors' estimation, because they attempted to alter scientific practice without affecting underlying power relations. In contrast, the most meaningful and effective of the transformative projects chronicled in this book altered the ways that scientists and engineers related to activists and community members, the ways that relatively powerless nonexperts were included in political decision making, and/or the ways that scientific data was connected to political action (e.g., Ottinger; Morello-Frosch et al.; Nieusma).

The concrete implications of this finding are not as simple to articulate or implement, yet they are essential to EJ advocates' efforts. The environmental justice movement already engages power relations. To take but one example, recognizing the relative powerlessness of poor communities of color against industrial facilities owned by large companies, EJ activists have organized multiple communities affected by one multinational corporation's pollution into an opposition movement that targets the corporation's global brand (see Doyle 2002). But activists' critical engagements with power do not always extend to their scientific activities: community-generated data are frequently inserted into expert-controlled regulatory processes, where it is readily dismissed. To make community-based science more effective, and hasten the transformation of scientific practices more generally, EJ activists need to begin to target the larger structures of power—including standardized practices and other institutionalized ideas of "good" and "bad" science (see Ottinger 2010)—that insulate science from participation by "laypeople." Our point is not that attention to power dynamics is new in EJ history, but that we need to enfold the processes of scientific and technological practices into the same conversations of structural change and empowerment.

Finally, the cases collected here show that established institutional structures, most often obstacles to change, can also provide resources for transformation. Frickel's chapter (chapter 1), for example, shows how scientific networks can play an important role in inspiring and fostering engagement between scientists and activists and in supporting technical professionals in the development of hybrid scientist-activist identities. Similarly, the energy engineers that Nieusma's chapter (chapter 5) focuses on were able to bring community voices to high-level policy discussions because of their unique, midlevel institutional location. The third recommendation of this book, then, is for experts themselves to look for ways to strategically employ their structural positions to broaden engagements between technical practice and environmental justice issues, and to help reconfigure power relations to give community groups more of a voice.

We conclude this book with the observation that science and technical practice are constantly in a process of transformation, that transformations in the direction of a more environmentally just science and technology are possible and are fostered by the engagement of technical practitioners with the EJ movement, and that transformative projects are made difficult by heterogeneous structural constraints. So it is that environmentally just transformations of technical practice are inevitably difficult, uneven, and partial—just as transformations of environmentally unjust incidents have been throughout decades of activism. If, as we suggest, change occurs as a result of disruptions in practice that provide scientists and engineers with new room to maneuver, the project of transformation becomes one of identifying, broadening, and benefiting from those ruptures. In such a project, not only EJ activists but also social scientific researchers and especially technical practitioners themselves are agents of change.

Notes

1. For clarity, we will refer to the now expansive field of STS through that acronym, not to suggest a unified theoretical or methodological approach in such literature but to point to a common umbrella term under which disparate and usually interdisciplinary scholars find common terms of reference and citation.

2. A number of researchers have posited a role for scholarship in transforming the scientific enterprise to make it more democratic, offering proposals for increasing citizen participation in policy decisions about science and technology (e.g., Fiorino 1990; Fischer 1999, 2000; Guston 1999), designing technology with the principles of democracy in mind (Sclove 1995, 2000; Nieusma 2004a), and using theory developed in the social studies of science to determine the relevance of different kinds of expertise to policy dilemmas (Collins and Evans 2002). Our approach differs from these in its emphasis on the race, class, and power disparities thematized by the environmental justice movement, which tend not to be represented in either the movements of scientists themselves or in STS proposals for democratizing science (though Nieusma 2004a is an exception). The environmentally just science and technology that we envision would incorporate the principles of sustainable and democratic science and technology while keeping equity issues central.

Part I

Forging Environmentally Just Expertise

1

Who Are the Experts of Environmental Health Justice?

Scott Frickel

Encomium

The environmental justice movement (EJM) lost a long-trusted ally when Marvin Legator died at his home in Galveston, Texas, in July 2005, and science lost a pioneering researcher and institutional trailblazer. In an obituary, Jonathan Ward, who until 2009 directed the University of Texas Medical Branch, Division of Environmental Toxicology (founded by Legator in 1976), describes his former colleague as "an unusual man," the characterization in reference to Legator's hybrid identity as an accomplished scientist and devoted activist ("Obituary for Marvin S. Legator" 2005).

Legator's accomplishments in science are clear. He held advanced degrees in biochemistry and microbial genetics from the University of Illinois Champaign–Urbana and put them to good use publishing more than 200 environmental health–related articles and books during a career spanning fifty-seven years. He was also a successful institution builder. As a senior research scientist at the Food and Drug Administration (FDA) in 1968, Legator went against the administrative grain of that agency to create the Laboratory of Genetic Toxicology—coining the name of a new environmental health field in the process (Wassom 1989). He was a charter member and executive officer of the Environmental Mutagen Society, founding editor of the journal *Teratogenesis, Carcinogenesis, and Mutagenesis*, and founding director of the Division of Environmental Toxicology (UTMB), a position he held from 1976 to 1999. But for Jonathan Ward and others who knew and worked with him, Legator's unique character stemmed from a deep commitment to social justice that infused and guided his scientific work. The general goals of Legator's research program, as posted on his laboratory webpage, spoke directly to this symbiosis:

Using the techniques of genetic toxicology and molecular biology, [we] conduct studies that will characterize environmental agents that may be carcinogenic, mutagenic or teratogenic. *The overall goal is to modify or eliminate substances such as industrial compounds that pose a threat to human health.*

An additional major goal is to work with communities that are potential victims of chemical exposure. *Organizing communities, exposure analysis, evaluation of health effects, and working toward a just solution are all part of this project.* (http://www.utmb.edu/pmch/Divisions/envtox/Legator/Legator.htm; my emphasis)

Good to his word, Legator was known in environmental circles in Texas and beyond as a "tireless champion" for industrial workers and for members of low-income communities that bear the heavy toxic burdens of industrial production (Environmental Mutagen Society 2005). Industry defense lawyers would probably characterize his work on the witness stand differently, but they would almost certainly agree with opposing counsel that Legator believed in the rightness of his political convictions as deeply as he believed in the integrity of his science. To the end, Legator seemed most at home on the uncertain cultural terrain created by the collision of grassroots environmental action and cutting-edge science. The dual arc of Legator's career cuts against the common wisdom that in choosing an activist path, scientists jeopardize their professional authority (see Bocking 2004). But was Legator as "expert activist" really that unusual? And if he was, what about his role made him so? What broader lessons might we draw from his life's work?

Answers to such questions are difficult to come by. Despite a rapidly growing body of research that investigates environmental justice protest and activism, few studies to date place professional scientists at the center of analysis (Allen 2003). What we do know is derived largely from case studies of environmental justice conflicts that pit underresourced community groups against powerful industrial and government actors (Lerner 2005; Sze 2007; Corburn 2005). Some attention has been given to the roles that individual scientists have played as activists in particular conflicts or settings, but researchers have yet to pay systematic empirical attention to the structure and dynamics of expert activism in the EJM. Our understanding of who these expert activists are, where they come from, and what difference their work makes to the movement and to science remains undertheorized and empirically underexamined. If nothing else, Legator's professional life and work—and the movement he fought for so passionately—encourage renewed critical attention to the topic. In that spirit, and consonant with recent calls to rejuvenate the political sociology of science (Frickel and Moore 2006; Moore et al., forthcoming), this chapter argues for an institutional approach to the study of expert activism in EJM. By institution I refer to relatively obdurate practices and meanings that are organized around socially recognized activities—such as knowledge production or social protest—and that shape human experience in ways that are often unrecognized. In Cohen and Ottinger's terms in the introduction to this book, experts frequently encounter institutions as "obstacles" against which they must work to create "room to move." But where their discussion of obstacles and opportunities places considerable emphasis on the role of individual experts and locally situated practices, I see institutions as broadly social accomplishments, interactively generated over time and space through social networks and conventions. This distinction is often lost in extant studies of expert activism in the cause of environmental justice.

By discounting experts' social locations within professional and political networks, existing research leaves a pressing gap in our understanding of how and why expert activism takes the forms it does. That gap is problematic because it limits scholarly appreciation for the scope and complexity of the phenomenon under study. Ignoring the gap also forestalls recognition of the transformative potential of expert networks—transformations that include generating organizational opportunities for advancing environmental justice. I argue that understanding how expert networks operate provides insight into the institutional conditions that promote and legitimate emerging forms of expert activism. Following a critical review of the relevant literature, I describe a research strategy for identifying networks of EJM expert activists and present data on experts' careers and political mobilization that illustrates the conceptual and political utility of an institutional approach.

Experts and Knowledge Politics in EJM

The rights-based discourse adopted by EJM community activists locates movement claims squarely within U.S. civil rights law (Cole and Foster 2001). Within that broad legal structure, however, community-level outcomes often turn on the technical merits of activists' claims regarding the existence of environmental contamination and the various impacts those hazards have on individuals and communities. The question of impacts, and particularly health impacts, depends further on the generation of credible evidence regarding chemical transport and bioavailability, the length, frequency, and routes of exposure, and a host of other factors. Thus, while the legal system sets the stage for framing and prosecuting EJM demands, the movement is deeply entwined by a politics of knowledge: who conducts environmental and health-effects research, how and where that research is organized, and who obtains access to it are common concerns. Toxicologists, epidemiologists, geneticists, physicians, and other health experts can influence outcomes in grassroots conflicts by reconfiguring the production and circulation of strategic forms of scientific knowledge.

This makes EJM's relationship to environmental health and allied sciences complex. On the one hand, the movement presents a broad challenge to the social authority of scientific knowledge (Couch and Kroll-Smith 1997). A variety of community-based research strategies developed within the movement to identify environmental hazards, document their deleterious effects, and mobilize that evidence in pursuit of movement demands illustrate this challenge most explicitly (e.g., O'Rourke and Macey 2003). On the other hand, the outcomes of local struggles are often dependent on the regulatory regimes that structure environmental assessment and on the knowledge practices and disciplinary commitments that feed mainstream environmental health research. Even when community groups are successful in generating their own data, the efficacy of their demands will likely depend on how closely their technical claims conform to the expectations of environmental health professionals and the assessment standards set by regulatory bodies (Allen 2004; Howard 2004; Ottinger 2010).

The significance of EJM's structural dependence on mainstream science is sharpened further by the gulf separating many at-risk communities from the professional community of environmental health experts. Systemic structural segregation marked by race, class, education, or occupational differences means that scientists and community activists do not typically live in the same neighborhoods, have access to similar economic and educational opportunities and resources, or even hold common expectations of political representation and procedural justice. But the gap is cultural as well as structural. "Rather than adopting the role of teacher or educator," writes Fischer (2000, 7, 9), "experts have largely set themselves off from the mass citizenry. Instead of facilitating democracy, they have mainly given shape to a more technocratic form of decision making, far more elitist than democratic" and one that can result in "the increasing unwillingness of citizens to accept uncritically the trained judgments of the experts." Such mutual mistrust can undermine broader EJM goals. To date, "collaborative efforts with . . . academics have been unsuccessful in expanding the movement's social capital to amplify political pressure for the significant structural changes necessary to facilitate a more equitable distribution of the environmental costs of production" (Cable, Mix, and Hastings 2005, 72). Because social inequalities and cultural expectations structured by race, class, education, and occupation combine to minimize interaction and maximize mistrust between environmentally burdened communities and environmental health professionals, scientists and engineers allied with the goals and strategies of EJM represent crucial conduits linking two vastly dissimilar social worlds.

Their importance to the movement is reinforced by narrative accounts that portray these expert activists as exceptional individuals. Case studies that include attention to community-based expert activism tend to celebrate those "oppositional professionals . . . who are willing to support communities in their efforts to seek recognition and assistance from industry or government" (Brown, Kroll-Smith, and Gunter 2000, 19). Expert activists who join with aggrieved communities typically face strong resistance from "government agencies and [other] professionals [who] often side with business in actively opposing connections between health and the environment" (p. 22). Because going against the grain of professional expectation can result in occupational sanction and damaged reputations, EJM activism is not a path that experts seem to choose willingly. In spite of these constraints—or perhaps because of them—expert-activists are credited with taking on a number of critical functions within EJM. These include translating technical information into language nonexperts can more easily understand; assisting in the design of community-oriented research projects; analyzing and organizing community-generated data; and representing community interests in court, the media, town council meetings, and other political and educational forums. In these ways and more, the success or failure of community efforts to seek and attain redress for environmental injustice often seems to hinge on expert activists' skillful provision of technical knowledge.

For all its appeal, this portrait of expert activism as an exceptional path that lone scientists pursue heroically at great personal and professional risk rests on relatively thin conceptual and empirical ground. The notion that expert activism is undertaken primarily by lone individuals largely ignores research on scientist activism that falls outside EJM. Much of that research stresses the importance of professional and political organizations, collective action framing, and alternative support and recruitment networks among experts engaged in various forms of protest (Delborne 2008; Frickel 2004a, 2004b; Martin 2006; Moore 2008;

Morello-Frosch et al. 2006). Additionally, by casting government, industry, and scientific opposition to expert activism as more or less monolithic, the exceptionalist account fails to recognize institutional opportunities that might facilitate interaction among state, industry, university, and social movement actors. These include the emergence in recent decades of government programs that have provided technical assistance to at-risk communities, funded community-based environmental justice research projects, and included community members on peer review panels; "good neighbor agreements" negotiated between industrial facilities managers and community groups; and the rise of a network of private foundations that encourage and support community-based environmental justice organizing. Perhaps most importantly, the empirical foundation of the exceptionalist account is built almost entirely from case studies. The overreliance on a single method introduces bias by highlighting the activities of a select number of highly committed experts whose involvement in particular organizations or episodes of struggle render them visible to case-study researchers. Such bias can distort resulting claims about the social nature of expert activism and forestall investigation of broader dimensions of what is an inherently complex phenomenon. Moving this line of research forward methodologically and theoretically requires a different approach.

Toward an Institutional Approach

An institutional approach is one that examines "the rules and routines, organizations, and resource distributions that shape knowledge production systems" and how those systems in turn condition society's uptake of scientific knowledge (Frickel and Moore 2006, 7). In the present study, institutional analysis locates expert activism among actors representing the state, universities, industry, communities, and social movements. It investigates how experts' social locations, networks, and associated professional experiences support and constrain coordinated responses to environmental injustice. This approach shifts the focus of research away from the isolated actions of individual expert activists and toward the organizational arrangements and network configurations that shape the mobilization of environmental science and health experts into EJM.

A key challenge this approach brings with it is the systematic identification of expert activists. While some experts are readily identifiable in the published literature or on EJM websites (and typically do not shy away from the "activist" label), many others are not so easy to find, for a variety of reasons. Some are understandably wary of gaining the spotlight of public controversy. Some are mainly involved in advising EJM organizers, work that takes place "backstage" at some remove from the front lines of struggle. Some make contributions to the movement via networking or outreach activities, forms of activism that are not straightforwardly contentious. Still others engage EJM sporadically, the visibility of their activism constrained by its temporal discontinuity.

The result is that the boundary-spanning networks connecting expert activists to EJM, to the state and other social institutions, and to one another tend to be largely hidden from public view in the form of "shadow mobilizations" (Frickel 2010). From this perspective, what is unique about the expert activists documented in extant case studies is not their activism per se, but the public visibility of their activism. I argue that these experts' public activism represents the tip of an iceberg, the submerged dimensions of which remain little understood. A deeper understanding of how social movements and professional knowledge workers reciprocally affect institutional change in science and engineering domains will be advanced by bringing the hidden experts of EJM further into the light.

To counter the difficulty posed by the relative public invisibility of these expert activists, I use nonrepresentative reputational sampling methods. I initiated research by interviewing organizers of three environmental justice support groups, two located in southern Louisiana and one based in New York City. These organizers provided me with names and contact information for my initial list of expert activists. The first stage of research resulted in thirty-two semistructured in-depth interviews with scientists, engineers, public health specialists, medical professionals, and environmental justice organizers conducted between 2002 and 2006. The twenty-four experts in this sample were all actively engaged in EJM at the time of the interview and the eight organizers all had active ties with multiple experts. While this is a small and nonrepresentative sample, as I describe below, the interviewees make up a highly diverse grouping. Viewed in network terms, the considerable variation in careers, areas of expertise, and institutional location among them offers preliminary insight into key structural features of EJM's shadow mobilization of expert activists.

The interviews ranged in duration from forty-five minutes to four hours, with most lasting one to two hours. Each interview covered set topics that included questions about family background, education, and employment history. Interviews with experts focused on their research, their involvement in EJM, and the connections they draw between research and activism. Interviews with organizers focused on organization history, goals, and strategies—including the tactical roles played by experts and the organizational challenges of connecting experts to communities in need.

I concluded each interview with a request for the names of additional experts. None of the thirty-two interviewees provided fewer than three new contacts (one provided twenty). This cumulative list of still-to-be-interviewed expert activists now consists of well over 100 names and contains surprisingly few duplicates. Clearly, much work remains. At present, I offer the list as provisional evidence of the networked character of this shadow mobilization. In the following two sections I present data from interviews with the expert activists identified in the first stage of research (N = 24), focusing on their professional characteristics and mobilization experiences.

Expert Activists' Professional Characteristics

Analysis of experts' educational backgrounds and employment records can provide a means of assessing the types of institutional resources that expert activists provisionally bring to EJM. The data summarized here demonstrate considerable variability along six key dimensions: type and number of advanced degrees, status of degree-granting institution, disciplinary training, academic rank, sector of employment, and employment trajectory.

By definition, experts are credentialed with postgraduate degrees, but how many, from where and in what combinations? Nearly half of the expert activists in this study (46 percent) hold PhDs or the medical school equivalent (DSc). Another 20 percent hold medical or law degrees and 80 percent hold master's degrees (MS, MA, or MPH). The credential-heavy character of the sample is expected, since for many practicing scientists and engineers the road to a PhD involves obtaining a master's degree in a related field along the way. Marvin Legator's educational background, which included a master's degree in biochemistry en route to a PhD in microbial genetics, illustrates what for many is a standard MS/PhD training sequence. But while about half of the expert activists interviewed for this study hold multiple advanced degrees, not all multiple degree holders followed the same educational path taken by Legator. Several expert activists hold two master's degrees in different fields—for example, in international development and environmental health. Two

MDs took second (master's) degrees in public health. One expert holds three advanced degrees—an MS in mechanical engineering, a PhD in nuclear engineering, and a second PhD in political science. Moreover, the large majority of the postgraduate degrees earned by expert activists in this sample came from prestigious universities, among them Cornell, George Washington University, Harvard, MIT, Rockefeller University, and the University of California–Berkeley. In short, data on degrees earned and degree-granting institutions suggests there is considerable depth in not only the quantity but also the provisional quality of expert activists' professional training.

That professional training is broad as well as deep, as indicated by expert activists' chosen field of study, academic rank, and sector of employment. While the U.S. environmental justice movements' tendency to focus on toxics and public health clearly shapes the constellation of expertise in this sample, the disciplinary training varies widely, especially at the PhD level. Of the eleven expert activists in the sample with PhDs, three of those degrees are in social and policy sciences, three are in basic life sciences (e.g., molecular biology), three are in medical sciences (e.g., anatomy), and one each is in physics and engineering. More strikingly, those eleven PhD degrees originate in ten distinct disciplines.

Academic rank is somewhat less varied. Among the expert activists employed at universities ($N = 10$), five hold the rank of full professor while four are employed in nontenured lines as research scientists or postdoctoral fellows funded through external grants or visiting professors on fixed teaching contracts. Only one expert activist interviewed for this study is employed as an untenured—and thus highly vulnerable—assistant professor. In other words, in this sample expert activists working in the academy are concentrated in positions where employment (if not credibility) is protected institutionally, or in positions where economically and symbolically there is relatively less to lose.

Employment sector is another important dimension of variation among these expert activists. Although they are heavily concentrated in the academy (eleven hold primary employment positions in universities or research hospitals), other major employment sectors are also represented. Two expert activists hold positions in federal government, six others work for nonprofit organizations, and one is employed in the research wing of a major chemical corporation. The four remaining expert activists also occupy positions in the private sector, either as owners or employees of small businesses or as physicians in private practice. The level of diversity in employment setting raises the related question of these expert activists' records of employment prior to occupying their current positions.

The data on employment trajectories suggests that relatively few took a linear path from graduate school into a postdoc and from there built careers in academics, government, or industry. Instead, even in this small sample, there is considerable movement by experts across employment sectors over the course of their careers. Legator's employment trajectory is a case in point. His first two jobs after obtaining a PhD were with chemical pesticide companies. From the private sector, he next took a position in government, heading a testing laboratory at the FDA before gravitating to academics. Others move in the opposite direction, beginning in academics and later moving into positions in government, industry, or nongovernmental organizations. For still others, experience in the nonprofit sector spurs an interest in academic or government research or program administration. The stories these data tell are complicated, but it is clear that the large majority of expert activists in this sample have not spent the bulk of their careers employed in one sector or another. Rather, they shift across sectors with some apparent regularity.

This brief data summary provides a number of preliminary insights about the types and diversity of institutional resources that experts may provisionally bring to EJM. For now, I address just three. First, high education levels combined with the status of the universities and programs from which many received training goes some way toward countering the notion—widely held in some scientific and regulatory arenas and often reinforced in the media—that the rigor and demands of more traditional scientific pursuits push less capable researchers toward community-based activism (see Mazur 1998; Brown and Clapp 2002). Judging from the data at hand, these expert activists are no less qualified than their peers whose research and public service activities hew closer to purportedly apolitical professional expectations.

Second, as a group these expert activists demonstrate wide diversity in disciplinary training, degree combinations, and accompanying skills. Assessed at the individual level, that diversity may prove a liability in situations where expert skill sets do not match up with the needs of specific organizations or communities. Viewed from a network perspective, the diversity of training and skills represents a menu of complementary resources that make possible the development of productive synergies among expert activists. The movement's ability to capitalize on that latent potential will depend in part on the extent to which expert activists are organized.

Third, employment mobility within this sample describes a "revolving door" of experts moving between academic, government, industry, and NGO jobs. Experts who work in different institutional settings over the course of their careers obtain a wealth of organizational experience that can be invaluable to EJM. All else equal, a toxicologist who knows what matters to regulatory scientists because she has worked as a regulatory scientist in the past will be more valuable to communities who have to deal with regulatory scientists in their quest for environmental justice than a toxicologist lacking that institutional experience. But in addition, the personal networks these experts develop through revolving-door employment afford them access to decision-making and knowledge production arenas that few other activists in EJM enjoy. While this dynamic has yet to be fully explored, employment mobility has important strategic implications for incubating an activist culture in and across multiple institutional settings.

Thus, in addition to the individual provision of technical skills that is emphasized in existing studies of expert activism, institutions of higher education and employment provide additional resources that experts can mobilize on behalf of EJM. Most significantly, these include the generation of boundary-spanning networks that cross knowledge fields and work settings to tie experts to one another in diverse ways. Such networks intersect with EJM, exponentially increasing the number and quality of resources that in principle are available to the movement. I believe there is great potential for these professional networks to transform knowledge politics and practice in environmental health and justice, *but only if experts are organized to do so.* In light of this largely untapped potential, we need also to investigate how those partially submerged networks become politicized and yoked into action for EJM.

Institutional Context and Expert Mobilization

Clearly activists in science are made, not born. But how and under what conditions? Interviewees suggested several different precipitating factors and social contexts that influenced their involvement in EJM, each providing a break or rupture of some kind to their conventional thinking. This section explores four: parental social values and family background, classroom experiences in college or graduate school, insights from research and professional practice, and direct experience with grassroots political struggle.

Parental Social Values

Several expert activists acknowledged the importance of their parents' social values and family background as precipitating factors shaping their own activism. "Very activist" is how one interviewee described parents deeply involved in civil rights work in New York City during the 1970s. Another noted a strong tradition in her family of helping others, of being "involved in the social aspects of the community," and linked that model to her current work with community groups. In neither case, however, was science an explicit part of those childhood experiences. But a third expert activist spoke of how science and politics merged in her family: "What got me to environmental health was, I was raised by a dad who's a scientist and a biologist, but also a naturalist. He's kind of a renaissance man. And a mother who was a politician at the local level. So I had these two pieces going on, and I really think a lot about that. It was kind of the scientific pursuit of knowledge but also politics."[1]

More commonly, interviewees described their parents' social values and their own upbringing as generally liberal but not specifically political or activist. Many also did not attribute much weight to the lessons gleaned during childhood in shaping their decisions as adults to engage EJM. Familial social values may have instilled in these future expert activists a strong sense of moral responsibility, but for most, their activism seems to have been triggered by something more specific and immediate, usually some tangible experience that connected environmental risks and social inequality more directly to scientific practices and institutions.

Classroom Experience

For many, those points of personal connection to abstract notions of environmental and social justice took place in the classroom. For one expert activist, politicization into EJM grew from an undergraduate communications course that introduced students to the Union Carbide disaster in Bhopal, India, as a case of poorly managed corporate public relations:

And the nature of the case study was . . . well, if you were the CEO for Union Carbide at the time, what would you have done differently to make it better. And I'm thinking, "Well, you killed thousands of people in one night. You can't make it better." . . . My interest in doing non-profit and social justice work really sort of grew out of this education that I had that showed me how, you know, the people that were pursuing my kind of career were being recruited by these big public relations firms to improve the image of companies such as Union Carbide that weren't doing a very good job of protecting the public health.

For another, participation in a master's program "framed around environmental advocacy" was the principal context for mobilization. "So, it was really during my master's [that] I started doing work with groups like Greenpeace, a local land preservation group in [state], and the Teamsters that really got me realizing, you know, who needed me more—industry or the advocates?" While this person's politicization was a direct result of an educational environment that encouraged advocacy, most graduate programs and faculty are not overtly aligned with environmental and social justice. More typical in this sample were expert activists who experienced a profound disconnect in graduate school between the professional expectations of the graduate faculty and social protest swirling outside the lecture halls and laboratories. As one expert activist who entered medical school in the late 1960s reminisced,

I started swimming upstream in medical school. That's when I decided . . . that medical school was not interested in community health or social justice. It was interested in world class research, in having Nobel prizewinners on their faculty and having Nobel prizewinners come to colloquia, which is all fine, top science and all that. But there's more to it than that, I felt. And because the political activism in [the city] at the time impressed that upon me: you can be both an excellent scientist and an excellent provider of relevant service and you don't have to sacrifice one for the other.

These examples describe experts' politicization as involving an ideological process of identifying abstract connections between expert knowledge and environmental inequalities.

Research Practice
For some, though, the connections are forged through more mundane processes. One expert activist explained his involvement in grassroots air monitoring growing from a meeting he had as a graduate student with a community activist in need of equipment capable of measuring sulfur dioxide emissions from a nearby refinery. The citizen's request had been rebuffed by university faculty and so the interviewee "called up the same [university faculty] people . . . and asked them about getting my hands on an SO_2 monitor for research, [and was told], 'Oh yeah, we've got like seven of them just mothballed in here, and, you know, you can take them whenever.'" In this example, the spur to activism is not grounded in ideological insight so much as the simple realization that, as this interviewee observed, "I just noticed that . . . with air quality . . . there's a lot of small things that could be done differently." As this example suggests, expert activism need not involve overt opposition to mainstream

science or science policy. In many cases, the most effective forms of expert activism may involve setting quietly transformative changes in motion (Frickel 2004). Examples of such "low-dose" activism also highlight ways in which the politicization of expert activists is shaped by professional practice.

For Legator, the thalidomide crisis that broke in 1962 was a key event that made visible the limits of genetic understanding and was instrumental in his efforts to build the country's first laboratory of genetic toxicology at the FDA. "As I went down the list of toxicological risks," he recalled, "it seemed that cancer was covered by our two-year bioassay. With thalidomide . . . at least we were aware of teratology. But the one thing that we were totally ignorant of, was mutagenicity. . . we had no way of truly identifying, in a systematic manner, compounds that were mutagenic." The thalidomide problem also revealed to Legator the importance of making genetic knowledge available to the wider public. As researchers in his FDA laboratory began identifying mutagenic chemicals, "It occurred to me that, what we really have to do to start helping people, that, it's great for us to publish our papers and get them into journals, but unless we do something more than that, nothing is going to happen."

Another expert activist who was a graduate student in the 1960s studied under scientists who had identified the biochemical mechanisms underlying thalidomide teratogenicity. Working in this milieu had a profound impact for this person. "The idea of environmental factors affecting development or human health was really brand new. It was radical, it was interesting to me." Another scientist this expert activist worked for had done groundbreaking work on gynecological diseases and had published a "very precautionary report about the profligate use of estrogen in oral contraceptives and the cattle industry; since we didn't know much about estrogen at all, introducing enormous amounts of exposures would be, in his opinion, [something] we should really worry about." Of these two, he observed "I guess . . . the two people who were my mentors and friends were both people who worked in the area of basic science with human health applications and who were willing to go against the grain scientifically." He concluded, "I was trained by two people who probably never thought of themselves as activists, but clearly both of them went against the grain, their science was in pursuit of a social aim, perhaps."

Social Protest

If the rupture of technoscientific controversies like the thalidomide disaster gave new meaning to otherwise mundane laboratory research, other expert activists were mobilized into EJM through the new spaces opened up by direct experience with grassroots political struggle. Three people interviewed for this study worked in antitoxics campaigns in the late 1970s and early 1980s. One of these movement veterans said that "[early] experience in a way changed my life. It changed my priorities, it changed the demands on my time, my interests. It focused them in a way that hadn't been focused before." His experience in community-driven hazardous waste cleanup efforts

"just made me realize that people who live in these communities don't have someone they can turn to—scientists who they can get honest information from, get honest answers. So it was pretty apparent to me that there was a big need for that kind of person or that kind of organization that could provide that [expertise], because all they have to turn to is government and universities, and universities weren't very much help."

This experience was a key "turning point" for him as an activist. Others describe a more gradual mobilization process. For some, this occurred during graduate school, when coursework and campus politics coincided in particular ways. For example, one university scientist described his teaching experiences as a graduate student during the late 1960s:

So I was in pharmacology giving lectures and as a graduate student we were assigned the worst lectures, the ones the faculty didn't want to give. . . . So I was given this job of giving lectures to medical students on the topic of toxic gases. And that was typically benzene, chloroform, and all of these things. And so I ended up doing a lot of research and giving a lecture on tear gas. And I became this guru of tear gas treatment, because all of the medical students were putting on yellow arm bands and going out and taking care of other students who were in protest marches and so they had to know about tear gas. So I guess it was a time where it was the people who I worked with were themselves, not radicals, but were socially involved. And then I was in a period of time and in a city where you really had stark choices. And not everybody was involved in the anti-war movement, but you couldn't, I mean, the city burned down in 1968, so it wasn't something you could ignore. So I guess in some ways, from my own experience, there were all of these things that kind of came together.

Others were radicalized before graduate school and that political consciousness shaped their course of graduate study. For example, one person describes working after college with an NGO involved in food aid issues in Central America. "It was through that work that I kind of learned public health on the ground," she observed. "You're working in the communities and you're flying in all of this food, but you're not dealing with the fact that communities don't have primary health care."

That realization was instructive in this person's decision to pursue a MPH in community health.

Still others become engaged in response to requests from local individuals or groups for information about specific environmental health concerns:

> I got a call from a man . . . who's wife was dying of cancer, whose father was dying of cancer. . . . And they had had a well drilled in their back yard and they feared that the well was contaminated. So I did one sample of his well. And through word of mouth, additional people would want their water sampled. And I [began testing for] four heavy metals. I charged them $100 for collection and analysis, and promised them that I would tell the [regulatory] agency if I found things so that we could get the agency to come in and do a more extensive analysis. So I did over a hundred wells, just through word of mouth.

As these later excerpts suggest, for many expert activists, getting involved in EJM was not based on a set of abstract concerns of moral or social responsibility, but on tangible experiences that in different ways connected social and environmental inequalities to science.

Network Diversity and Depth

The excerpts in the preceding section demonstrate a diversity of experience, context, and motive for "swimming upstream" as expert activists. In different ways, the recognition of resource disparities has led to the politicization of experts and to their mobilization in EJM. For some, science represents a way to find answers people need. For others, science—or more specifically the distribution of knowledge interests within science—is part of the problem. As echoed in this book's introduction and the other chapters, in response to those concerns some have taken up high-profile roles within the movement; others have used their experience and professional capital to effect change more subtly in EJM communities and in professional and policy arenas.

Taken together, these examples help illustrate the ways an institutional approach can broaden understanding of expert activism. Rather than maintain a relatively narrow focus on expert activism as it emerges within particular EJM organizations or conflicts, this study considers how the politicization of experts' understanding of science and their mobilization into EJM are shaped by networks and institutional location. The result is a more analytically complex picture of expert mobilization processes. While some interviewees described mobilization as coming directly from experience in community struggle, for many others politicization processes began earlier in their careers, in the context of their education and training in graduate or medical school. This finding suggests that expert activists are made in the classroom and at the laboratory bench as well as in the streets (see Ottinger, chapter 9, this volume).

An institutional approach can also deepen understanding of the forms expert activism takes. For example, by taking into consideration contexts other than the front lines of grassroots conflict, we find that expert activism in EJM goes well beyond the provision of technical expertise. Institutional analysis helps demonstrate how resource disparities between universities and EJ communities provide opportunities to experts who take advantage of their institutional locations to bridge those gaps, either by linking those communities to underutilized resources such as mothballed air-monitoring equipment, or by reframing the ways expert knowledge is made meaningful, as exemplified by Legator's repeated efforts to redirect knowledge production in his laboratory and department toward the needs of environmentally burdened neighborhoods.

Conclusion

In this chapter I have argued that networks structure and coordinate interactions among experts and EJM. Those networks tend to be highly dynamic, bridge multiple institutional settings, and operate below the radar of public view as shadow mobilizations. Studying shadow mobilizations requires a shift in analytical focus toward an institutional perspective, from individual experts to expert networks—a move that marks important conceptual and empirical steps forward.

Findings from this preliminary study challenge the heroic characterization of expert activists that figures so prominently in the dominant exceptionalist account. At minimum, the study suggests that there is no "one-size-fits-all" model of expert activism in EJM. Not surprisingly, expert activists are a diverse lot, bringing different attributes, experiences, skills, and resources to the movement from different institutional locations. They find different ways of engaging in the EJM, and in doing so face different types of risks, and influence different sorts of outcomes. Rather than discount the significance of these multidimensional differences as simple contingencies of history, locale, or personal biography, an institutional perspective contextualizes individual action, finding untapped opportunities in the diversity of experts tied to one another through political and professional networks.

Similarly, the institutions that expert activism challenges in the name of EJM are not so universally constraining as to negate all but the most heroic efforts. Clearly, as this book theorizes, there is "room to maneuver" in science, where a range of opportunities exist that activists working inside science and government and even industry can exploit to productive effect. For individuals, opportunities may emerge from one's familiarity with the rules guiding regulatory processes or the informal access one has to scarce resources controlled by a university department. Yet a far broader set of opportunities emerge through the generation of expert activist networks that crosscut scientific, medical, and engineering fields. These findings have implications that are political as well as academic.

For EJM organizers and community activists, shadow mobilizations represent a broad and largely untapped well of technical and organizational resources. As such, the occasional advantage gained by activists in recruiting a sympathetic expert may not lie in the technical skills that the expert is able to wield in a specific struggle, so much as in the social capital she brings to the movement in the form of network ties to other experts with different and complementary skill sets and organizational resources. All else equal, organizers who maintain connections to ten to thirty experts, rather than only one to three, greatly increase their potential for achieving more continuous expert mobilization and that stability in turn can alter fundamentally the organizational calculus of grassroots struggle. The challenge for EJM organizers is to develop ways of identifying the scope and structure of these networks so as to most effectively nurture and grow the opportunities embedded within shadow mobilizations.

These same networks can also provide for the coordination of resources, interests, and action in institutional domains of research and decision making typically far removed from the communities that form the front lines of EJM. In theory, shadow mobilizations can operate as mechanisms for channeling local knowledge from EJ communities into professional domains and for incubating activist cultures inside science and engineering. To capitalize on that potential, experts working in EJM must utilize these networks strategically. This will not be simple because many expert activists I have spoken with do not seem aware of the networks that might connect them to like-minded peers. As one scientist told me, "No one else is doing this." This sense of political isolation, while fairly common among those interviewed for this project, is also clearly inaccurate. But it speaks to the submerged nature of shadow mobilizations and to the low-dose character of much expert activism such that it remains invisible even to other expert activists. Thus an additional challenge for EJM experts and organizers alike will be to find ways to better organize expert activists without sacrificing the political advantages gained by their relative invisibility.

Additional research on shadow mobilizations can help activists, organizers, and experts better understand how professional and political networks entwine in EJM. To date, these dynamics have received limited attention from science and technology studies and social movement scholars. As a result, the capacity for such loosely bounded networks to significantly transform expert practices, participation, and identities, and thus alter EJM knowledge politics, is an empirical question that remains largely unexamined.

One way forward is with research that moves beyond case studies to systematically investigate shadow mobilizations and the institutions of science, education, and politics that condition their emergence. Such research needs to critically examine the "shadow" aspect of shadow mobilizations. That is, why does the political potential represented by professional networks intersecting EJM remain largely latent and underutilized? Formal organizations likely play an important role. What sorts of changes in the culture and organizational structure of science would make university departments and professional associations and conferences more effective contexts for tactical innovation and expert recruitment into EJM? As other chapters in this book demonstrate, cultures of activism in science are also likely to be instrumental in suspending shadow mobilizations at the margins of public visibility and recognition. How do the institutionalization of scientific practices and the construction of new audiences for environmental health expertise shape the form, level, and duration of expert activism in EJM? To the extent that empirical research guided by an institutional framework provides more accurate understandings of expert activism and its coordination through a shadow mobilization of submerged networks, researchers stand not only to inform academic understanding of knowledge politics in EJM, but also to play a more direct interventionist role in changing it.

Notes

1. I have removed all names and identifying markers contained in the interview transcripts to protect the anonymity of people quoted in this chapter.

2

From Science-Based Legal Advocacy to Community Organizing: Opportunities and Obstacles to Transforming Patterns of Expertise and Access

Karen Hoffman

In the mid-1970s, a group of concerned citizens formed an environmental organization, the Clean Air and Water Network, to watchdog the implementation and enforcement of the then-new Clean Air Act, Clean Water Act, and other environmental laws, in a large metropolitan area of the United States.[1] The organization was small, consisting of a few scientists and lawyers. Their project—keeping tabs on the regulatory system as it was in formation—was massive and socially and technically complex. It was also challenging in that it attempted to change social relations that allowed industries to evade government regulations intended to prevent pollution—relations that have existed since the beginning of industrialization, and that have been fundamental in industrial society.

In the mid-1980s, the Clean Air and Water Network was inspired by the emergent environmental justice movement to broaden its activities beyond science-based legal advocacy. The Network was persuaded by the analysis of the environmental justice movement that showed that in addition to the injustice against the public in general of industries polluting the air and the water, there is an additional injustice in patterns of pollution: economically and racially marginalized people bear a disproportionate burden of pollution and, in fact, are targeted by developers for the siting of new sources of pollution.[2] Motivated by this analysis, the Clean Air and Water Network branched out from science-based legal advocacy into community organizing in areas particularly burdened with pollution. In doing so, the Network intended to share and proliferate its expertise among people living in particularly burdened areas, and thus enable them to access and participate in regulatory decision making. This added project, too, was challenging in that it attempted to change social relations that block the access of many people—especially those with few resources and those who are otherwise marginalized—to ways of gaining the knowledge and skills needed to participate in and influence political decision making.

In the language of this book, the Clean Air and Water Network's encounter with the environmental justice movement created a rupture. This rupture created an opportunity for the Network, as it saw it, to sharpen its analysis of what was wrong in the world and what needed attention, and to expand its mission and practice according to its assessment. Although the Clean Air and Water Network has grown since its beginning in the mid-1970s, it remains a small organization, with few employees and no profit stream. Hence we should expect that the organization made some movement toward accomplishing its second enormous and radical project, altering patterns of who participates and influences political decision making, via community organizing, but that it also encountered some constraints. This chapter examines this movement and these constraints, and suggests some ways of working to overcome the constraints.

The analysis presented here is a close examination of one campaign within the Clean Air and Water Network: the Dioxin Prevention Campaign, which worked to prevent dioxin pollution by identifying sources of this potent toxin in the region, pushing government to improve its regulation of dioxin, and pushing industries to make materials substitutions or process changes that would allow these facilities to avoid producing dioxin.[3] The analysis is based on eighteen months of participant observation and interviews in 1999 and 2000, in the Clean Air and Water Network, primarily focused on the Dioxin Prevention Campaign and its relations with other environmental and public health advocacy organizations, regulatory agencies, and neighborhoods that are particularly burdened by pollution. Beyond the particular case, the chapter offers insight into emergent phenomena with implications for environmental justice advocacy more broadly. The case is remarkable in that the Network was one of the few environmental groups that made such big changes, or in fact any changes at all, in response to the environmental justice movement's critique. Additionally, because many environmental organizations feared being charged with elitism and racism by EJ critics (Open Letter 1990a, 1990b), gaining access to study environmental responses to the environmental justice movement was difficult; the analysis presented here thus offers unique insight into a process of change instigated by environmental justice activism. In the forefront, at the time of the study, of activities by environmental organizations to respond to EJ critics, the Network's experience suggests practical and structural obstacles to change that organizations undertaking comparable changes may be expected to encounter.

The Clean Air and Water Network's Encounter with the Environmental Justice Movement

When the Clean Air and Water Network decided in the mid-1980s to expand into community organizing, the organization's activities had for the ten years of its previous existence consisted of investigating and identifying instances in which federal and state pollution control laws were not being enforced or implemented in the region, and using empirical data and legal tools to push for enforcement and implementation (Max Girardeau, interview by author, February 28, 2003). In the mid-1970s, a round of federal pollution control laws that had a much greater capacity for enforcement than previous laws—including the Clean Air Act of 1970 and the Clean Water Act of 1972—were being put into practice (Kubasek and Silverman 1999). Before this round of laws, a regulatory system barely existed; government did very little regulation of pollution (Esposito 1970; Zwick and Benstock 1971; Miller 1987). The new capacity for pollution regulation came, in large part, through the provision for citizen suits (Miller 1987). That is, the Clean Air Act of 1970 and every other major environmental law that followed gave standing to individuals, allowing them to sue the government on behalf of the public for not implementing the law or not implementing it properly, and industry for not following the law. With this round of laws, government could accomplish much more regulation if the concerned public was monitoring whether or not, and how well or poorly, the regulatory agencies were implementing the law, and bringing or threatening to bring lawsuits when the regulator fell short.[4]

The then-new regulatory system consisted of the federal Environmental Protection Agency (EPA) and state and regional agencies that carried out EPA regulations. Its job of implementing the pollution control laws was enormous, and made even larger by the resistance to regulation that came from polluters. The job, granted to concerned citizens, of monitoring the implementation by regulatory agencies was equally enormous. Concerned citizens organized to do this work began forming nonprofit environmental organizations. The Clean Air and Water Network was one such organization.

Hence the work of the Clean Air and Water Network from the early 1970s to the mid-1980s consisted of scrutinizing the regulatory system, and identifying and pushing it to fulfill its unfulfilled mission. Network scientists worked at identifying undone regulation, by accessing and analyzing regulatory documents that industries were required to produce under the new laws. For example, the scientists analyzed documents that reported on pollution being emitted from industrial facilities and on compliance with the new laws and regulations. After the scientists identified undone regulation, Network lawyers pushed—using the threat of lawsuits or actual lawsuits—government and industry to get that regulation done. Frequently, industry fought these legal challenges, employing scientists to dispute the interpretation of data on which the cases were based. The Network's scientists did the work of supporting the empirical basis of the cases in these struggles (Max Girardeau, interview by author, February 28, 2003). In other words, when it undertook the new work of community organizing, the Clean Air and Water Network was engaged in legal advocacy based on investigative and scientific work. In this chapter, I call this work "science-based legal advocacy." This project was massive, complex, and against the grain of two centuries of industrial development with little or no effective government regulation of pollution.

In the mid-1980s, the Clean Air and Water Network's activities brought it into contact with other pollution prevention organizations, including some new (at the time) organizations concerned not only with the injustice of industries harming the public in general through polluting the air and water, but also with particular trends newly identified as unjust: the fact that economically and racially marginalized areas receive a larger burden of pollution than other areas, and are targeted by developers for the siting of new polluting facilities. These new organizations' awareness of and motivation by the social injustice of patterns of pollution was distinct from the awareness and motivation of the Clean Air and Water Network and other environmental organizations that had existed up until that time.

The organizations with this distinct awareness of environmental injustice with which the Clean Air and Water Network interacted included, among others, the United Church of Christ Commission for Racial Justice and the National Toxics Campaign Fund (Max Girardeau, interview by author, February 28, 2003).[5] These are some of the organizations that contributed to the formation of the environmental justice movement. Like the science-based legal advocacy of environmental organizations that came before them, their strategies of fighting environmental injustice included identifying local instances of undone or weak regulation, bringing legal challenges, and gathering evidence and making interpretations of that evidence that support the legal cases. But the environmental justice organizations were also different from the earlier environmental organizations in important ways. One thing that was distinct about environmental justice organizations was that they recognized that frequently the people living in areas with many pollution sources, or targeted with new pollution sources, were not involved in the environmental review of projects proposed for their areas, nor other implementation of environmental regulations (Open Letter 1990a, 1990b; NTCF 1993). To address this problem, environmental justice organizations worked with people living in particularly burdened areas to develop pollution prevention campaigns through community organizing, rather than hiring one or two scientists and lawyers to develop and carry out these efforts on behalf of those communities.[6]

Additionally, environmental justice organizations developed an ethic of having community organizers work with residents of particularly burdened areas in a way that shares with them the expertise—the knowledge and skills—that is needed to advocate for themselves, rather than having community organizers, or scientific, technical, and legal experts, work on their behalf.[7] Hence, in addition to organizing with the residents of particularly burdened areas, what was distinct about the new environmental justice organizations, compared to environmental groups like the Network, was their aim to empower the residents.

So compelling were the analysis and strategies of the environmental justice movement to the Clean Air and Water Network—the awareness of environmental injustice, the need to broaden the effort to include people living in particularly burdened areas, and the goal of doing so in a way that shared with them the expertise needed to advocate for themselves—that the organization expanded its activities to include community organizing in addition to science-based legal advocacy. As Network researcher Jennifer Heath put it, the staff intended "to provide the community [that is, people living in particularly burdened areas] with resources to be able to do the work themselves," and to help them "develop a community base, . . . to develop leadership capacities, . . . research capacities, all kinds of different capacities" (interview by author, June 20, 2000).

The environmental justice analysis created a rupture in the analysis and practice of the Clean Air and Water Network. The rupture was not one that brought a halt to the organization's activities; rather, it inspired the Network to expand its practice. The break did not so much create an opportunity for new directions, in the sense of providing resources for new work, as much as it motivated the organization to make the considerable effort involved in expanding its practice.

The Clean Air and Water Network's activities when it expanded into community organizing were aimed at addressing and changing a structural inequality in our society, one in which some people have the expertise needed to enter and participate in the space of political decision making while others do not. This structural inequality manifests in participation in environmental decision making, but is far broader than this area of social life. This structure is produced and reproduced in environmental policymaking, but other institutions, particularly those of schooling and work, are the pillars of its production and reproduction.[8] Through our participation in these institutions, we get divided and divide ourselves into expertise-haves and expertise-have-nots.[9] This uneven structure often shields political and legal decision making from most people—from all expertise-have-nots, and particularly those with few resources and those who are otherwise marginalized. In other words, it shields political decision making from democracy. When environmental justice organizations engage in community organizing, they are attempting to make an end run around this uneven structure.

Of the scholarship on expertise and participation in political decision making, most has focused on the production of expertise on the margins, by laypersons, those cast as expertise-have-nots, who manage to become expertise-haves (Epstein, 1996; Di Chiro, 1997). In this analysis, I am focusing on a different area: how expertise is pervasively and unevenly produced and distributed, how the Clean Air and Water Network attempted to undo this uneven structure, and what the constraints in reaching this goal were. In doing so, I illuminate opportunities for transformation that were realized and those that were not.

New Practices: Movement and Constraint in the Clean Air and Water Network's Expansion

In the Environmental Organization
Having decided to expand its activities to include community organizing in addition to science-based legal advocacy, but having no experience with this new activity, the Clean Air and Water Network took a cue from other organizations that specialize in this area, and hired another kind of worker—community organizers—to do this work. In this particular endeavor and in the overall project of attempting to share its expertise

with people in particularly burdened areas via community organizing, the Clean Air and Water Network both made movement toward its goal and encountered significant obstacles. For example, the staff successfully raised funds to hire community organizers, but not always on the schedule needed. This was the case with a campaign to prevent dioxin, begun by the Network in 1994 and built without a community organizer through 1998 (Carol Jensen, interview by author, July 27, 1999).

The Dioxin Prevention Campaign's mission was to identify dioxin sources in the region and to push offending industries to make materials substitutions or process changes that would allow those facilities to avoid producing dioxin. Mark Zilles, a scientist with ten years experience in the Clean Air and Water Network, began building the campaign in 1994.[10] He investigated what dioxin source testing, if any, had been done in the region. He urged the regional Department of Air Quality and Department of Water Quality to inventory potential dioxin sources and to test whether they were actual sources. Having identified air and water sources, Zilles developed a plan for phasing out dioxin-producing materials and processes at all sources. To help along the phaseout he was advocating, he pinpointed as best he could the origin of dioxin in each of the sources.

In the four years before hiring an organizer, Zilles also worked at assembling support in the region and the state for the Dioxin Prevention Campaign. About forty-five organizations involved in environmental and health activism signed on as part of the statewide "Alliance for Zero Dioxin." Zilles worked with contacts he already had, and cultivated new ones.

The knowledge and ability to do the activities that Zilles did between 1994 and 1998 reflect the kinds of expertise that the Clean Air and Water Network had to offer newcomers from particularly burdened communities. But neither newcomers nor community organizers were involved in those years. Thus, the decision to do community organizing moved the Clean Air and Water Network toward its goal, but the inability to immediately fund new community organizer positions presented an obstacle.

In 1998, the Network hired community organizer Carol Jensen to work on the Dioxin Prevention Campaign. This achievement moved the Clean Air and Water Network toward its goal of sharing its expertise with people in particularly burdened areas. Yet Jensen encountered significant obstacles to the work of community organizing (Carol Jensen, interview by author, July 27, 1999). The Dioxin Prevention Campaign focused on the seven petroleum refineries in the region, particularly one in the town of Springfield that was known to be violating its dioxin permit limit and that would be the subject of an upcoming hearing at the regional Department of Water Quality. Jensen's job was to do community organizing in the neighborhoods near the Springfield refinery, and to take some of Zilles's work of maintaining and continuing to build the Alliance for Zero Dioxin. Initially, the Alliance building took so much time that there was none left over for building alliances in the refinery neighborhoods. The high volume of work in contrast to the small number of staff in the Clean Air and Water Network got in the way of community organizing.

Jensen, furthermore, found herself in a climate in which the old style of working—having a scientist build a campaign and identify support in other existing environmental and health organizations—continued to be the reference point from which many of the Clean Air and Water Network staff and board members drew their expectations, despite their openness to the new ways of working.[11] This meant that in addition to community organizing, a large part of Jensen's work was to educate the rest of the organization about what community organizing is, what its intended outcomes are, the timeline on which it gets done, what resources are needed, what counts as an accomplishment (if not winning a lawsuit, or publishing an impressive investigative report), what obstacles come up, and the limits of an organizer's role and duties. Thus, hiring Jensen furthered the organization's plans to share its expertise with residents of particularly burdened areas. But negotiating what her work would be, and whether and how the Network would allow her to expand its practice, needed to be done alongside of, and took time away from, reaching out to and working with newcomers.

In Particularly Burdened Neighborhoods

Early on, Jensen realized that the high workload required an additional organizer. She was able to hire one on the condition that the person would, in addition to community organizing, raise funds, including her own salary. This condition inherently limited the effectiveness of the new employee, but was a fact of life in the Clean Air and Water Network. Nine months after the Network hired Jensen, she and Erin Whittaker, the new organizer, headed out to the refinery neighborhoods. They made presentations at neighborhood council meetings throughout the city on the Dioxin Prevention Campaign, covering the health effects of dioxin exposure, industrial sources, and the fact that dioxin production is preventable; they also collected names and contact information of people who wanted to get involved in the campaign. Within six months, the organizers quite readily connected with many concerned residents, who signed up to become involved in the campaign. Meeting people in particularly burdened neighborhoods, and interesting and involving them in the campaign, were not obstacles.

Space for Newcomers

Despite the interest in the Dioxin Prevention Campaign that Jensen and Whittaker found in the refinery neighborhoods, the relations that emerged between these newcomers, the community organizers, the scientist, and the lawyers created obstacles to the Network's goal of sharing its expertise. After Jensen and Whittaker assembled a critical mass of about twenty refinery neighbors, they held a larger meeting in a community center in Springfield in order to introduce all of the newcomers to each other and to Zilles.[12] At this meeting, the organizers also reviewed and facilitated discussions about the information on dioxin exposure and its health effects. Additionally, Zilles presented the campaign strategy he had already developed.

Much work still needed to be done to see the plan through—including lobbying local and federal regulators for their support for dioxin prevention, keeping abreast of a dioxin risk assessment that was in progress at the EPA, providing investigative support to the Network lawyers in suing the refinery for violating its permit limit, and working with the media to get the campaign publicized. Each of these activities required forms of expertise that newcomers to environmental advocacy did not have. These activities offered opportunities for sharing expertise with less experienced peers, and for learning from more experienced peers. But no mention was made of this remaining work at the meeting with the Springfield residents. Over the coming months, Zilles alone worked on these activities.

The one place Zilles saw a role for the refinery neighbors was with their attendance and show of support for the Network's position at the hearing on the Springfield refinery's dioxin permit limit at the regional Department of Water Quality.[13] These observations beg the question of why Zilles envisioned such a limited role for refinery neighbors. I address this question below, continuing here to narrate the subsequent events.

Several refinery neighbors gladly obliged Zilles. Working with the community organizers in a series of later meetings, Zilles prepared the refinery neighbors for the hearing by telling them that the refinery was violating its permit limit, and that its managers were asking the

Department of Water Quality to allow the violation, pleading that they could do nothing more to reduce dioxin. Zilles reported that the refinery managers' claim was suspect and that it reflected a common strategy among polluters of resisting regulation, but that the Department was leaning toward the refinery managers. The job of the Clean Air and Water Network and its allies was to warn the Department against this and, rather, to press the Department to enforce the permit limit and to pressure the oil company to investigate ways of eliminating dioxin.

What happened at the hearing fit Zilles's description very well; however, it took place in language nearly incomprehensible to anyone but highly experienced practitioners. The newcomers—myself included—could not understand the dialog. For example, the engineer reporting the Department of Water Quality's position gave the following reason for allowing the refinery to continue exceeding its dioxin permit limit, at least for a period of up to one year (or, in the engineer's language, "proposing a Cease and Desist Order, or CDO, for an amendment for the refinery that is going to extend the final compliance date by up to one year"):

[The refinery's] current violations (of its water pollution permit) appear to be caused by atmospheric deposition. . . . So the issue is, is [the refinery] the primary source of that air deposition? . . . A primary cause of [the refinery's] permit violations was atmospheric deposition from dispersed sources. This was because of the somewhat unique way [the refinery] manages its wastewater. [The refinery has a] two-mile long discharge canal. . . . The top end is where the discharge occurs to the Bay, and that's also where the permit limit applies. [The refinery] uses large ponds and canals to treat and manage its wastewater. They combine treated process wastewaters with other waste streams like storm water prior to discharge to the Bay. The discharge from the treatment plant would be in compliance with the permit limit, but after mixing with the other waste streams, the combined flow violates the limit. All of these waste streams are impacted by atmospheric deposition simply because of the large size of the area that drains to them. For instance, the coke pond and non-processed storm water area, they're each about 70 to 80 acres in size. (from meeting transcript)[14]

The newcomers had no opportunity to read or otherwise be informed of this analysis prior to the hearing.

Nonetheless, through this experience the refinery neighbors gained exposure to a part of the space of participation—the Department of Water Quality, a government regulatory agency—as well as its events, procedures, authorities, and the structure of authority there. These newcomers were also exposed to the highly experienced practitioners and the specialized language and persuasive rhetoric they used to argue their cases. The event presented questions for the newcomers that could be the starting point of acquiring the expertise needed to advocate in the regulatory arena. But there were no opportunities within the Clean Air and Water Network for the newcomers to follow up on those questions. No room was made within the Network for newcomers to gain expertise through contact with their more experienced peers.[15]

Hence the Network tried to develop this pollution prevention campaign together with residents of the particularly burdened community of Springfield in a way that empowered them, yet at the same time the process and results were limited.

Understanding the Constraints: Knowledge, Time, Resources

The Constraints

Given the openness and interest among the staff of the Clean Air and Water Network to share their expertise with newcomers from refinery neighborhoods, and given that the organization took significant steps toward this goal, why did Zilles not create opportunities for the newcomers from refinery neighborhoods to become involved in the activities of the campaign? Why did the community organizers not do anything more with respect to creating space for the newcomers to become involved? These are crucial questions for understanding the campaign's movement toward its goal and its constraints. Interestingly, these questions were on no one's mind, not even mine, while I was doing my fieldwork. In the flow of everyday life in the environmental organization, these questions were in the realm of that which no one asks. There was a division of labor in which the highly experienced scientist continued to do the work of developing and carrying out the pollution prevention campaign, the newcomer community organizers connected with residents of particularly burdened areas and built support for the campaign among them, and the newcomer refinery neighbors became known to the campaign staff (and vice versa) and showed their support for the campaign. That division of labor was unquestioned and relatively unable to be questioned.

Zilles did not create more opportunities for newcomers for two reasons. First, he did not know that newcomers need these opportunities in order to learn.[16] This was in part because facilitating learning had never been a part of his job. He had no experience in helping people acquire the skills and knowledge that he has. It was because of the lack of knowledge on the part of the scientists about working with newcomers from particularly burdened areas that the Clean Air and Water Network hired community organizers.

Second, Zilles did not more fully involve newcomers in his work because he did not have time: his workload already overburdened him.[17] As can be seen in the above description of Zilles's activities in the Dioxin Prevention Campaign, his responsibilities included identifying undone regulation, investigating what was known and not known about the problems he identified, and seeking contact with regulatory agency officials to persuade them to create undone regulation and to produce the knowledge that was needed in order to do so. Where neither government nor industry had any plans for addressing a problem (in this case, dioxin prevention), Zilles developed a plan. This work involved learning about the industries to understand where the problem originates—a task that is complex and time consuming in terms of both accessing the information and mastering the technical dimensions of the operations. Zilles also worked at assembling support for the campaigns, by networking with other activists and organizations. As the Clean Air and Water Network added community organizing, he also needed to attend events being held in the communities, which frequently took place at night and on weekends. In addition to his work in the Dioxin Prevention Campaign, Zilles supported multiple legal cases that the Network's lawyers were evaluating and/or carrying out. He also fielded calls and requests from concerned citizens about all manner of things—from a worry about a pile of railroad ties being burned next door, to a need for speakers regarding a shareholder initiative for pollution prevention at an annual meeting of a polluter in the region.

Not only is funding a constant obstacle for grassroots organizations; so too are time and human resources. The additional demands for technical expertise when addressing industrially produced environmental risks such as dioxin only exacerbate those problems. Working with newcomers takes time and Zilles had no spare time. The fact that scientists and, indeed, everyone employed by the Clean Air and Water Network, were overcommitted in their workload and unable to take on the new project of community organizing was another reason that the Network hired community organizers.

Given that the community organizers Jensen and Whittaker had knowledge of how to work with newcomers to environmental advocacy and how to facilitate their acquisition of skills and knowledge, and given that they were hired specifically to do community organizing, why did they not create opportunities for the newcomers to get involved in the work? Why did they not work with Zilles to create these opportunities? Initially, Jensen could not begin any aspect of community organizing because of the amount of overflow from Zilles's workload she was expected to take over. That is, networking with potential supporters and collaborators in the building of the Alliance for Zero Dioxin took a great deal of her time. After Jensen solved this problem, she and Whittaker encountered a more difficult but less obvious or even perceptible obstacle: the idea of creating opportunities for newcomers to get involved in the activities of building and carrying out the campaign was outside of the discourse and practice of the Dioxin Prevention Campaign and the larger Clean Air and Water Network.

Jensen's and Whittaker's method of community organizing was to work together with residents of particularly burdened communities to identify problems, analyze problems, and devise and work toward solutions. But in the division of labor in the Dioxin Prevention Campaign, only a narrow aspect of working toward solutions—showing up at hearings and other public events to demonstrate support—was available to the newcomers. Jensen and Whittaker did not work with Zilles to create opportunities to get involved in the work of the campaign because this slot for newcomers in the division of labor was relatively unable to be questioned: it was not an option.

Addressing the Constraints

Zilles's lack of awareness—and, indeed, that of others in similar positions inside and outside of the Network—that newcomers need opportunities to work together with their more experienced peers in order to acquire skills and knowledge, could have been addressed by letting him know what was required. For example, Jensen and Whittaker could develop a training session for Zilles and/or others on this need.

Zilles's time constraints were a result of the Network's "economy": it had an enormous and complex project, and relatively little and insecure funding to support the work. Its project was to figure out what regulation of pollution was incomplete or insufficient and either to push government agents in their region to create the necessary regulation, to push industry to change its processes or materials, or both. At the same time, the Clean Air and Water Network's "economy" had relatively few resources for accomplishing its project. As a nonprofit organization, the Network was not in the business of selling products or services and, therefore, had no profit stream. It relied on voluntary grants from philanthropic foundations and donations from individual members. Between the early 1970s and the mid-1980s, the Network was able to employ one lawyer and several scientists to do the work of critically monitoring the adequacy of air and water pollution permits for an entire large metropolitan area and advocating for more adequate regulation when needed. At the time of the research, the organization was able to support no more than five lawyers and three scientists.

These conditions of an enormous and complex project and limited resources produced the constraints on Zilles's time to create opportunities for newcomers to become involved in his work. At times, these constraints became more severe, because the workload of Clean Air and Water Network was linked in important ways to the number of instances in which industries were not complying with regulations, government was not enforcing regulations or has weakly implemented regulations, and industries were resisting the making of new or more comprehensive regulations. When there were many items needing attention, the few resources of the Network were stretched even thinner.

The organization's workload was also linked to the amount of resistance from industrial firms. The Network frequently found itself involved in cases in which industries resisted the regulation that the organization pushed for. Large, relatively wealthy firms, such as the oil refiners in the region that the Network watchdogged, have the financial capacity to put up significant resistance. Frequently, this kind of activity quickly diminished the organization's already limited resources. As scientist Jennifer Heath put it, "It's really hard to staff. We have a small staff compared to how much [the large corporate polluters] spend on these cases. It's a tiny fraction of what they're spending. . . . For us, to fund a couple of staff people for two years, before you get your costs covered, is very difficult" (Jennifer Heath, interview by author, June 10, 2000). Industry resistance to regulation thus created even more work for the Network's small staff, including pursuing additional lawsuits, seeing through lengthy lawsuits, and giving input to regulatory agencies to counter the input of polluters.

Thus, because of the economy of the Network (and, indeed, many nonprofit public interest organizations), Zilles's time was always scarce, and the actions—and inaction—of polluters or regulators made his time even more scarce. Additional demands on Zilles's time, such as those that community organizers might make, threatened his ability effectively to manage his already too-heavy workload.

This brings us to the question of how to address the problem of community organizers not creating opportunities for newcomers to get involved in the work of the campaign because doing so is outside of the discourse and practice of their organization. Changing the discourse and practice of an organization, or even knowing that some alteration is needed, is difficult because members go about their work inside this discourse and practice. A broader perspective is frequently beyond reach. Jensen and Whittaker—themselves newcomers to the Network, bringing a new practice to the enterprise of science-based legal advocacy—clearly needed more support in order to fully register and speak up about this need. This is an area in which the work of an informed outsider, such as an ethnographer, in this case, can identify the limits of the discourse and practice of the organization, and the need to expand them, and thus provide the support community organizers needed. Jensen and Whittaker have long moved on from the Clean Air and Water Network, but other community organizers remain in their complex situation of needing to work with, but to avoid being eclipsed by, scientists and lawyers doing science-based legal advocacy. An outsider participating in and studying their practice can, either in the moment or after the fact, feed back to them their perceptions about this need and thus support them in developing an awareness of the lack of space in the existing discourse and practice of the organization for what they were hired to do. This in turn would expand their ability to bring their own needs as community organizers, and the needs of communities they work with, to the attention of the Campaign and the Network.

It would also be useful for the Network and other science-based legal advocacy organizations that attempt to expand into community organizing to maintain awareness that they have initiated an expansion and hired community organizers to carry it out, and to expect that the expansion they initiated will likely create some changes in their own ways of working. Scientists, lawyers, and other advocates in this situation would further their own intentions by being open to change, being aware of their resistances, and giving attention to how to protect and maintain their practices and projects, while at the same time helping the new project of community organizing to flourish.

Conclusion: Escaping the Double Bind

I return here to the problem with which I began this chapter: that expertise is required for entry into the arena of regulatory decision making, and many people lack this expertise because expertise is unevenly produced in our society. The Clean Air and Water Network attempted to transform this pattern by doing an end run around the way expertise is unevenly produced in institutions of schooling and work. They did so by reaching out to and working with refinery neighbors in a way that shared with them, and proliferated among them, knowledge and skills needed to access and influence regulatory decision making.

In the case of the Dioxin Prevention Campaign, the Network both made progress toward this goal and bumped into limits. Part of the progress was having the openness to expand their community and their practice to share knowledge, skills, and experience with newcomers from particularly burdened neighborhoods. The progress also consisted of developing the financial capacity needed in order to hire community organizers. The latter was a significant achievement for a relatively small nonprofit organization that relied on donations and had a limited budget.

The community organizers quickly and easily moved the Network forward by finding, connecting with, and beginning to build relationships with refinery neighbors who shared the organization's concerns about dioxin and other pollution. This was also a notable accomplishment for a small group of experts who were accustomed to working only with each other.

The Dioxin Prevention Campaign hit limits in trying to proliferate its expertise after connecting with the newcomers and inviting them to show their support for a campaign that was already developed before they had become involved. At this point, the campaign provided no further way for the newcomers to get involved in carrying out this project, and thus to learn from their peers who had the expertise needed to access and to advocate in the regulatory arena.

These limits surfaced because the Clean Air and Water Network undertook the expansion from working among a small group of experts to trying to share its expertise with others as a possible way out of extremely difficult circumstances. That is, the organization wanted to do all that needed to be done to make the regulatory system function in ways members thought was just—including the newly embraced task of working with residents of particularly burdened areas in a way that empowered them. But the limits on the Network's resources made this impossible, at least in the short term. They placed the organization in a double bind, in the sense characterized by Wilden and Wilson as "a situation that requires a choice between two states which are equally valued and so equally insufficient that a self-perpetuating oscillation is engendered by an act of choice between them" (as quoted in Fortun 2001, 363). In this case, the Network found its original practice of policing the regulatory system insufficient, and moved toward expanding into a second project, trying to proliferate expertise and to increase advocacy capacity through community organizing. But the Network went only part way before encountering obstacles to that plan and swinging back mainly to limit its work to its original project.

Double binds cannot be escaped by choosing one of the unsatisfactory options presented. To transcend a double bind, one (or a group of people) must recognize the circumstances as impossible, and respond to the impossible circumstances in some way. The case of the Clean Air and Water Network showed that in-depth analysis of circumstances is a difficult task for nonprofit organizations attempting to accomplish multiple profound societal changes with limited funding, time, and human capital. Through this ethnographic analysis, as an effort to complement the steps taken toward social change by the Clean Air and Water Network, I have identified four areas that need more attention in order to take further the expansion into community organizing and the effort to transform the uneven production of expertise, and that may help find the way out of the double bind.

One area has to do with knowledge. The Clean Air and Water Network needed more knowledge about how to facilitate the acquisition of expertise by newcomers. The second area has to do with resources: the community organizing project needed additional time from the scientist. Therefore the Network needed an additional scientist and, thus, funds in order to facilitate the acquisition of expertise by newcomers.[18] The third area is support for community organizers, who themselves are newcomers to the Network, bringing a new practice and trying to implement changes requested by the organization. They needed support to be able to register and speak up about the need for some time from the scientist and, thus, the way the scientists' practices need to change. The fourth and last area that needed attention was awareness among scientists and other staff members with a long history in the organization that their own practices might change, and attention to how to protect and maintain their practices and projects, while at the same time helping the new project to flourish.

More generally, ethnographic analysis of the Network's efforts to proliferate expertise highlights obstacles to EJ transformations of scientific practice at two levels. The case shows the significance of mundane factors in projects of transformation. Time, money, and other resources can create real constraints on scientists' (and others') ability to open up new room to maneuver, even when they set out to change their practices. But resource constraints exacerbated a second, deeper obstacle: unexamined assumptions and priorities that sustain the hierarchies that surround scientific knowledge and practice. When resource constraints created a tension between getting science-based legal advocacy done and inventing new ways to share technical expertise with expertise-have-nots, the scientists' work was allowed to take precedence over the new and unfamiliar work of organizing. The fact that, despite the Network's commitment to community organizing, neither scientists nor organizers thought to question this dynamic is evidence that the work of scientists has a privileged status, even among people who wish to disrupt the inequalities between expertise-haves and expertise-have-nots. The case thus suggests that taking full advantage of opportunities to transform science requires not only additional resources, but also an ability on the part of advocates from all backgrounds to understand how deep, naturalized assumptions subtly shape their priorities and their ultimate effectiveness—an ability that sustained engagement with critical but sympathetic social analysts can help to foster.

Notes

1. The Clean Air and Water Network and, except where noted otherwise, all names of persons and places in the text are pseudonyms. I sometimes refer to the Clean Air and Water Network as "the Network" in the text below.

2. The U.S. environmental justice movement works to transform socially unjust uses of the environment, understood as air, water, soil, and land. A large part of this movement is constituted by locale-based groups that organize—frequently, with support from each other—simultaneously against the siting of hazardous and polluting facilities in areas that already host their share of such facilities, and against a pattern in which racially marginalized and poor communities are particularly burdened with pollution sources and hazardous industries or waste facilities and often are targeted for the siting of these locally unwanted land uses (Bullard 1993; Cole and Foster 2001).

3. *Dioxin* refers to a group of a persistent organic pollutants that bioaccumulate in the food chain. These toxins are highly potent carcinogens and also cause quite a few other serious health problems in extremely low doses. According to an EPA draft risk assessment circulated in 2000, average background levels of dioxins in humans in the United States are high enough to cause a lifetime cancer risk as high as one in a thousand. This level is a thousand times higher than the risk level that is generally considered "acceptable" by regulatory agencies (one in a million). Dioxins are created when organic matter and chlorine are together at very high temperatures. Prevention of dioxin has been achieved at some of the largest sources, such as pulp and paper mills, but even smaller sources are important and undesirable, because dioxins persist and bioaccumulate. The number one exposure is through eating common foods, especially those higher up on the food chain, because dioxin has been dispersed in air and water and then concentrated in the food chain.

4. There were additional elements of the new environmental laws that allowed for greater effectiveness in regulation. For example, they removed restrictions on who could participate in agency rulemaking, giving any member of the public this right (Bollier 1991). Additionally, the Clean Air Act of 1970 and the Clean Water Act of 1972 required the EPA to make federal standards and to derive from these standards limits on air and water pollutants at industrial facilities throughout the country. Earlier pollution control law had required states to create standards and pollution limits, but this law was ineffective because companies could avoid the standards by migrating to states with low standards. The new federal standards applied to the entire country and were intended to prevent this kind of migration.

5. These are the actual names of the organizations. The National Toxics Campaign Fund no longer exists.

6. Another aspect of environmental justice organizations that distinguishes them from the environmental organizations that came before them is their use of direct action, in addition to legal approaches.

7. As Di Chiro (1997) and Choy (2005) have discussed, what counts as expertise is socially and historically specific. In this case, the forms of expertise needed to access the courts and administrative agencies include having knowledge of the issues under debate in the regulatory arena, fluency in the technical language in which debate is conducted, knowledge of the law, and ability to find violations of the law and to support claims of violations with scientific evidence.

As Di Chiro (1997) has pointed out, laypersons have kinds of expertise that are relevant to the making of environmental regulations, but they often are overlooked by policymakers.

8. This structural inequality also manifests in most if not all areas of policymaking, as well as job getting, schooling, and other arenas of social life. Anthropologists and sociologists of learning (the way expertise is acquired), science (a main category of expertise), and social and cultural reproduction have shown us that we live in a society in which expertise is durably unevenly produced and distributed, along lines of race, class, gender, and other privilege, in institutions including, especially, schooling and work. Social scientists of education and social and cultural reproduction, such as Bourdieu (1977), Willis (1977), Eckert (1989), Holland and Eisenhart (1990), Fine (1991), Lave and Wenger (1991), Lave and McDermott (2002), McDermott (1997), and Wenger (1998) have shown how school success and receipt of credentials through schooling (traditionally prerequisites for acquiring expertise) are unevenly produced along the lines of class, race, and gender. Working within the anthropology of science, Traweek (1988) has described the acquisition of expertise in the training and work life of high-energy particle physicists, the social processes of selection and elimination of trainees according to the number of available positions (and not necessarily or primarily according to their degrees of expertise), and the ways these processes are tightly coupled with gender subjectivity.

9. It should be noted that whether we have or do not have expertise frequently has little to do with our capacity to acquire these forms of expertise.

10. The following account of Zilles's work on the Dioxin Prevention Campaign is based on a report he wrote and an interview with him on March 28, 2000.

11. This interpretation is my own and is based on conversations and experiences with Jensen during my fieldwork, and an interview with her on July 27, 1999.

12. Although, by this time, the Network employed five lawyers, several interns, and a legal assistant, none of the legal staff attended meetings with the refinery neighbors. The reason requires additional investigation.

13. My concern about the limited available roles for the newcomers from refinery neighborhoods stems from a belief that in order for people to gain expertise or, more simply, to learn, they need to be involved and have contact with practitioners who already have mastery over what is to be learned, and their practices. This view is informed by Lave and Wenger's (1991) discussion of learning as "legitimate peripheral participation." I address this concern later in the chapter.

14. The transcript continues:

[The Clean Air and Water Network] comments that the refinery's air emissions could be a significant source of the air deposition that's causing the discharge violation. [The Network] cites as one of the bases that data from a catalytic reformer at the [refinery] shows that this air source is about 25 times what would be allowed under our water discharge permit. We agree that this air emission is much greater, and we think it is more like 20 percent. But while some dioxins from this reformer air stack may deposit within the refinery and be captured by the storm water or the coke pond, the majority would transport to further distances. And, more importantly, looking at the congener group profiles of the air emission and the water discharge, the two are not similar.

It may be of interest to note that since the whole refinery shut down in March, the company has reported that their canal discharge dioxin concentrations have stayed essentially the same, compared to last year when the reformer and the refinery were in full operation.

It's also important to note that [the company] has stated the commitment to voluntarily reduce the emissions from the reformer. . . . The Air District has determined that the source does not pose a significant enough risk to warrant additional notification or control measures pursuant to AB-2588, which is the Air Toxic Hot Spot Information and Assessment Act. (meeting transcript)

15. It should be noted that the newcomers also did not follow up their experience at the Department of Water Quality with questions, or requests for opportunities to ask questions and debrief. This absence is part of the dynamic that reproduces the uneven distribution of expertise, but a full discussion of this topic is beyond the scope of this chapter.

My view that possibilities exist in a confusing experience for gaining expertise (learning) and my focus on whether, in this case, these possibilities were opened up or closed down by the more experienced practitioners in the Clean Air and Water Network are shaped by Lave and Wenger 1991 and Wenger 1998.

16. This statement reflects my interpretation of my experiences and observations of the Dioxin Prevention Campaign.

17. This statement reflects my own perspective on the Dioxin Prevention Campaign.

18. If the Network pursues the goal of raising funds to hire another scientist, this will bring them into competition with other environmental justice organizations for the finite funds that are available through philanthropic grants and member donations. This possible action thus may well create conflict with allies and take the Network into another double bind, for which they would need to search for a way out.

Additionally, if the Network attempted to fundraise for an additional scientist, it may not succeed, given its lack of a profit stream and dependency on voluntary contributions, and given the current difficult economic times. In this case, all the Network can do is understand that it is in a double bind and do the best it can do.

3
Toxic Transformations: Constructing Online Audiences for Environmental Justice

Jason Delborne and Wyatt Galusky

The emergence of web-based communication media devoted to the dissemination of data on pollution and risk to affected communities presents an opportunity to transform political contestations about environmental justice. In particular, governmental and nongovernmental organizations have leveraged the digital availability of the Toxics Release Inventory (TRI) database with web-based mechanisms to publicize comparatively large polluters. As Fung and O'Rourke (2000) have argued, the public availability and accessibility of TRI created pressure on polluters, often generated in the corporations themselves, to reduce emission numbers that were high relative to industrial competitors. This kind of "regulation by shaming," while not perfect, did spur some highly public reductions in self-reported toxic pollution (Graham 2000). It also generated some hopeful rhetoric touting the transformative power of the Internet in the realm of pollution—empowering people to take control of their lives, their communities, and their environments (see, for example, Foster, Fairley, and Mullin 1998; Gates 1999; Tuttle 2000). Web-based presentations of TRI data, spurred by and fostering new patterns of interaction among the public, information, and experts, created a rupture in technical practice that gave voice to the values of environmental justice.

The rhetoric of empowered citizenry and political transformation has long been a part of the Environmental Protection Agency's "TRI Explorer" and Green Media Toolshed's "Scorecard" websites, among others. These sites position themselves as empowerment tools precisely by providing what communities advocating for environmental justice have lacked—information on pollution and its effects (see Galusky 2004). This theme of information poverty, common to web-based forms of intervention, has dominated the historical development of the environmental justice (EJ) movement. Efforts to produce and disseminate this kind of data, while both laudable and necessary, have exposed a central challenge within the EJ movement: the integration of various forms of expertise and sources of power by connecting scientists and laypersons. As numerous chapters in this book demonstrate, these ongoing activities have generated meaningful engagements that address information poverty, but they can and do encounter difficulties in empowering specific constituencies (see, for example, Powell and Powell, chapter 6; Hoffman, chapter 2; Morello-Frosch et al., chapter 4; and Liévanos, London, and Sze, chapter 8, this volume).

The websites themselves, in emphasizing access to information as the main path to environmental justice, miss an opportunity for a broader transformation of political engagement. In this chapter, we aim to demonstrate this point in two steps. First, we argue for understanding these websites as modes of performance, which configure various roles and relationships that organize expertise and civic action. This analytical lens, developed below, foregrounds three kinds of design decisions that construct audiences for EJ science: a historical understanding of the problem of environmental justice, a choice about which particular publics should be involved, and a claim about what particular forms of participation should be available. Specifically, we reveal how a seemingly generic online database about toxic pollution (the TRI) does not operate as a "neutral" resource. Instead, choices in website design imply heterogeneous understandings of the obstacles to environmental justice and divergent potential roles for website visitors to participate in the movement.

Second, we show that this analytical frame reveals the extent to which the emphasis on information poverty in the design and implementation of web-based forms of intervention reinforces the "deficit model" of the relationship between scientists and the public (see, for example, Wynne 1996b; Irwin and Wynne 1996). Put simply, the deficit model defines the *problem* as a lack of understanding among laypersons, which prescribes a *solution* of experts transferring their knowledge. For reasons outlined below, this framework presents a continuing obstacle to broader opportunities for transformation of public involvement and presumed roles for action. At the same time, this study of website design furthers another point that threads through the present book, namely that the values and motivations of the environmental justice movement show evidence of having shaped new scientific and technological practices. Specifically, in the cases studied here we find that historical attention to information poverty has shaped the design, function, and identity of several toxics-related websites.

We begin with a brief overview of the model of audience construction (Delborne 2011) before providing a historical understanding of the EJ movement with specific attention to the theme of information poverty. These sections stage our analysis of three TRI-based websites, which we discuss in terms of how their designs construct audiences, including their expected competencies and prescribed roles. We conclude by discussing strategies for ensuring more effective and just distributions of TRI data and the broader implications for transforming participation by both laypersons and experts in the EJ movement as a whole.

Performing Science, Transforming Participation

The analytical framework we use to engage the transformative potential of EJ websites highlights how scientific information never exists apart from human action—it is always performed for an audience (e.g., the readers of a journal, peer reviewers, or portions of the wider public that consume mass media addressing scientific issues). Key here is the observation that the audiences of any particular scientific performance have a kind of irreducible power: (1) the performance only has impacts as it affects the behavior of its audience(s); (2) audiences *do* affect the content of the performance—they are never completely passive. We must then ask to whom is science being performed and to what end does the performance engage the audience? As we argue below, insights gleaned from theater studies can reveal vital, often implicit, components of EJ struggles to change policy and to change people.

By invoking the dramaturgical frame of performance to consider scientific practice in environmental justice struggles, we expand the typical focus on "scientific" performers to include attention to their audiences. Such audiences might include EJ activists, regulatory bureaucrats, residents of affected communities, polluters, and political representatives. As Bennett (1990, 167) relates, "A performance can activate a diversity of responses, but it is the audience which finally ascribes meaning and usefulness to any cultural product." Within the EJ realm, cultural products include scientific facts and specific knowledge about chemicals, exposure, risk, and appropriate responses.

Applying a framework developed by Delborne (2011), we are interested here in three specific and important aspects of the dramaturgical approach that apply to environmental justice controversies. First, the historical understandings of the problem and the people involved play a role in shaping the current presentation of any controversy. The historical legacy of past performances—how audiences have typically been engaged with the problem at hand—operates by offering preestablished audiences for future performances who may already know what to expect and to do. This "alreadyness" of the performance, importantly, can operate positively (in calling to arms those for whom the performance is staged) or negatively (by neglecting potential future audiences or by reinforcing actions that do not positively engage the audience so constituted). Historical continuity not only reveals but conceals, enables, and impedes.

Second, audience composition occurs in the context of explicit and implicit choices made about whom to include and whom to exclude from the performance, choices that have repercussions for social and scientific power. This audience composition issue is an important factor for any staging of a scientific controversy, particularly in terms of efforts to include and exclude certain audience members or to generate a particular audience profile. Who is invited to participate or to witness, either explicitly (e.g., through direct engagement) or implicitly (e.g., through choice of venue)? Audiences get composed through "extrascientific" choices of time, space, and staging, and thus can be seen as neither accidental nor arbitrary. Audience compositions also reveal the differences between planned inclusions/exclusions and real audience identities—that is, the intended versus the actual audience.

Finally, audiences are constructed under certain assumptions about the competencies of audience members and the roles they may play. In theater, an audience can be expected to have certain language skills, to possess shared cultural experiences, and often to sit patiently in fixed seats until appropriate moments for applause. For scientific audiences, assumed competency plays a significant role in how the science is performed. In one scenario, audiences surrounding environmental pollution may be assumed to possess all the necessary tools and social competencies for viable political engagement, minus the critical information offered by the performance. This "being told" places the audience in a dependent role with knowledge-producing institutions and fails to foster meaningful transformation. On the other hand, one might imagine a performance that encourages the audience to offer their own stories (and thus convey a kind of legitimacy or value on those stories) based on specific and local experiences—a model with strong connections to Boal's (1985) "theater of the oppressed." In exploring the transformations of scientific practice in environmental justice, we suggest that the specific, constructed character of such performances presupposes distinct types of interactions. Some of these empower laypersons to engage with technical information while others do little more than recreate historical patterns of dependence and separation.

A History of Information Poverty for Toxic Audiences

Focusing on audience construction allows analysts to interrogate how performances are structured. To whom are they performed? Why is there a performance at all? What problem is the performance designed to solve?

Considering these questions, we find the theme of *information poverty* weaving through several historical moments of environmental justice, a theme that has significance for understanding the politics of science about toxics. Specifically, concerns about deficits and the continued lack of scientific information (ignorance, to varying degrees) remain central to the performances of environmental justice reviewed here. The narratives imply that scientific information is the missing piece to the puzzle, that if such data could only be produced and provided, environmental justice would quickly result. We argue that an exclusive emphasis on providing information pushes aside other important political considerations, such as empowerment, means, maneuverability, and will. Narratives not so singularly focused on information poverty embody such concerns—for example, those that focus on the drivers and designs of production, the (dis)incentives for "green" innovation, the balance between private and government responsibility for safety testing, the consequences of designating production processes as trade secrets, and the political preference for margins of safety and tolerance of risk.

The contemporary legacy of current U.S. legislation on toxics can be traced to the incidents of Love Canal and Bhopal (see Szasz 1994; Fortun 2001). Earlier traces, however, can be seen in the industrial hygiene movements of the early twentieth century.[1] These movements viewed the hazards of the workplace as resulting not from corporate malfeasance or willful neglect, but by a lack of information regarding toxic substances associated with manufacture (see Nugent 1985; Rosner and Markowitz 1987; Gottleib 1993; Sklar 1995; Sellers 1997; Stradling 1999). For example, Alice Hamilton, one of the first scientists to systematically study occupational disease, found lack of knowledge to be responsible for most of the problems related to toxic workplaces: "Neither 'deliberate greed [n]or even actual indifference' lay behind the 'iniquitous conditions' she uncovered in industry after industry . . . ; rather, she placed most of the blame on 'ignorance and an indolent acceptance of things as they are'" (related in Sellers 1997, 73). Hamilton thus "set the stage" for future performances of information poverty as the key driver of toxic harm.

The problem of toxic exposure radiated outward, from a problem for workers in industrial settings to a problem for communities who had unknowingly found themselves exposed to toxic waste sites. The events of Love Canal did much to galvanize public attention to this new toxic waste dilemma (for detailed descriptions, see Levine 1982 and Gibbs 1982), and the story catalyzed the public imagination (Fortun 2001, 77). Importantly, the incident is perhaps best remembered as one where particular residents of Love Canal began to fight for increased safety and information about the threats to which they were exposed. Szasz (1994, 38–54) argues that Love Canal helped propel antitoxics as a "mass issue"—one taken seriously by citizens and government officials.

The problems of Love Canal were presumed to center around two distinct knowledge deficits: first, the fact of exposure (a lack of awareness of the presence of buried waste); second, the possible and real effects these chemicals were having on residents' health. The latter deficit proved the more troublesome. Existing scientific evidence of chemical toxicity, along with the presence of such chemicals in a particular area, did not provide conclusive evidence that residents had been unduly endangered. With the help of scientific researchers such as Beverly Paigen and external environmental groups including Environmental Defense, Lois Gibbs and the Love Canal Homeowners Association (LCHA) conducted their own toxicological surveys and formulated exposure rates and theories on the flow of toxics (Gibbs 1982; Levine 1982). The LCHA had limited success in generating politically viable scientific information, but they did manage to create enough data and attention to galvanize a public perception that something was wrong. In dramaturgical terms, the LCHA had to create a performance of the information they generated, but had little success with an audience of government scientists. Instead, in restaging for politicians (including President Jimmy Carter) and the media, the LCHA constructed an audience with the capacity to respond.

Love Canal has become part of the historical legacy of environmental justice and antitoxics activism. Its continual performance in policy circles legitimizes a national intervention strategy that reinforces the idea that more information will transform practices in favor of environmental justice. Unfortunately, this backgrounds the contentious process through which that lack of information was overcome and communicated, concealing its original communicative and dramaturgical context.

Coupled with other disasters, including Woburn[2] and Bhopal, a narrative of information poverty became institutionalized within national policy. How the chemical disaster in Bhopal is remembered offers a particularly strong example. Fortun (2001), in her study of reactions to the Bhopal tragedy by various interest groups (corporations, activist groups, governmental departments), notes the differences in the stories these groups tell regarding the incident: an unfortunate accident, an act of sabotage, yet another example of corporate malfeasance, or a consequence of information deficits. These groups then incorporate that act of remembrance into their organizational structure, establishing the context for future performances. For the U.S. government, the tragedy of Bhopal was due to a lack of knowledge about nearby chemicals, and this perspective became enshrined in legislation—specifically, the Superfund Amendments and Reauthorization Act (SARA), which in turn mandated TRI (see also Hadden 1989).

So emerged "Bhopal's Babies" (Fortun 2001, 63)—environmental regulations intended to facilitate the disclosure of risk. These pieces of legislation, which include SARA and TRI, solidified the understanding of the problem of toxics as one of a populace lacking the necessary information about toxic releases and their consequences. Consequently, environmental organizations have mobilized around this legislation, using the resultant database tool (the TRI) as a primary means of intervention, namely, providing citizens with information.

TRI Websites: Constructing Toxic Audiences

All of these previous performances form the backdrop in front of which current, digital versions of environmental justice activism are staged. We focus on three TRI websites: one maintained by the EPA itself and two operated by NGOs that rely on the EPA's data. The websites embody scientific performances and thus construct audiences both implicitly and explicitly. As new spaces of participation where citizens—as Internet users—interact with scientific data ostensibly to deliberate about environmental action, these websites thus represent attempts to transform the pursuit of environmental justice.

In compliance with the Emergency Planning and Community Right-to-Know Act (EPCRA), enacted by Congress in October 1986, the EPA began tracking industrial releases in computerized databases. The motivation for computerizing the information presumed that individuals could become more aware of toxic releases if they found out about them electronically. The EPA website declares:

> The goal of TRI is to empower citizens, through information, to hold companies and local governments accountable in terms of how toxic chemicals are managed. . . . The [TRI] data often spurs companies to focus on their chemical management practices since they are being measured and made public. In addition, the data serves as a rough indicator of environmental progress over time. (http://www.epa.gov/tri/triprogram/whatis.htm)

While this mission statement itself implicates a number of audiences (communities, citizens, companies, researchers), we examine the actual websites as performances that construct much more finely grained audiences with particular characteristics.

The three websites we analyze—EPA's "TRI Explorer," OMB Watch's "RTK NET," and Green Media Toolshed's "Scorecard"—all aspire to cater to the information-starved citizen. The sites entreat the user to discover more about their environment and to use that information to make a difference *in* that environment. The websites appeal to an audience interested in becoming "better" citizens and in improving the health and integrity of their community through informed action. These sites thus position themselves as empowerment tools and their audience members as effective users who gain power with access to information. Our analysis suggests that the websites fall short of such optimistic claims in numerous ways.

Particular elements of each site best illustrate the means by which the sites position their audiences vis-à-vis the knowledge being presented. Four general categories and questions follow the basic framework articulated above, examining how the homepage and deeper links position the audience, the data, the experts, and the political and social context:

- Historical articulation of the problem and means of intervention (Why is this site/data important? How has the website emerged within the context of environmental justice?)
- Composition (How is the audience composed spatially and socially given the website's mode of aggregation and use of the communication medium itself?)
- Presumed competencies (Who is included or excluded, and what are those who show up expected to know and understand in order to process the information?)
- Presumed roles (What is the audience expected to do during and after the performance?)

To varying degrees, each site answers these questions, through explicit description and implicit organization.

TRI Explorer (Version 4.7) (EPA)

The TRI database was first produced in 1988, with a specific mandate to be publicly accessible. The TRI homepage (see image 3.1) represents the contemporary web-based public access to the database (http://www. epa.gov/tri/). The site adopts a simple layout, providing information on what TRI is and why it exists. It also provides links to featured topics and program-related materials. Entering a zip code in the prominent box, labeled "Search TRI Data," brings the user to a new window with a chart, labeled "Releases: Facility Report" (see image 3.2). This chart details the amounts, in pounds, of chemicals emitted by the company listed in the far left column. Below each company is a list of the emitted chemicals they reported, as required by law. By clicking on a company name, the user can access a breakdown of the reported chemicals, including (after one more click) charts that track release trends over the reporting dates. A user can also find the address and contact information for

Figure 3.1
EPA TRI homepage screenshot

U.S. ENVIRONMENTAL PROTECTION AGENCY

TRI Explorer

Recent Additions | Contact Us   **Search:** ○ All EPA ⊙ This Area [＿＿＿＿＿＿] (Go)
You are here: EPA Home » TRI » TRI Explorer(ver 4.7) » Reports

Releases: Facility Report

* * * New Feature in This Release of TRI Explorer * * *
Detail columns are collapsed by default. Click the **◄▐►** icon to view additional columns. Use your Browser back feature to collapse.

Data source: Release Year 2006 PDR data set frozen on October 12, 2007 and released to the public February 21, 2008   See Note   **Go To New Report**

TRI On-site and Off-site Reported Disposed of or Otherwise Released (in pounds), for facilities in All Industries, for All Chemicals, zip code 13323 in New York, 2006

Row #	Facility	TRIF ID	Total On-site Disposal or Other Releases		Total Off-site Disposal or Other Releases		Total On- and Off-site Disposal or Other Releases
			◄▐►	◄▐►	◄▐►	◄▐►	
			▲ ▼		▲ ▼		▲ ▼
1	INDIUM CORP OF AMERICA, 36 ROBINSON RD, CLINTON	13323NDMCR23ROB	24		994		1,017
	CERTAIN GLYCOL ETHERS		1		0		1
	LEAD		23		994		1,016
2	JAMES H. RHODES & CO, 3683 STATE RT 12B, CLINTON	13323LCTRN3683S	9		0		9
	4,4'-METHYLENEBIS(2-CHLOROANILINE)		9		0		9
	Total		**3**		**32**	**994**	**1,026**

Back to top

Export this report to a text file ⓘ
Create comma-separated values, compatible with spreadsheet and databases.
(Download) all records

View other report type:
○ Transfers Off-site for Further Waste Management; or
○ Quantities of TRI Chemicals in Waste (waste management)

Note: Reporting year (RY) 2006 is the most recent TRI data available. Facilities reporting to TRI were required to submit RY 2006 data to EPA by July 1, 2007. TRI Explorer is using a *"frozen"* data set based on submissions as of October 12, 2007 and released to the public on February 21, 2008 for the years 1988 to 2006 (i.e., revisions submitted to EPA after this time are not reflected in TRI Explorer reports). TRI data may also be obtained through EPA Envirofacts

Off-site disposal or other releases include transfers sent to other TRI Facilities that reported the amount as on-site disposal or other release because not all states and/or not all industry sectors are included in this report.

Figure 3.2
EPA TRI facility report screenshot

the company. Clicking on a chemical name brings up specific reporting information for that chemical (including whether or not the chemical is a trade secret[3]), along with how the company interacts with the toxic substance (as manufacturer, processor, user) and if it is transported (searched May 2010).

RTK NET (OMB Watch)

RTK NET, run by OMB Watch (Office of Management and Budget Watch), has been in existence since 1989 (see image 3.3).[4] The website "provides free access to numerous environmental databases . . . [allowing users to] identify specific factories and their environmental effects, and assess the people and communities affected" (http://www.rtknet.org/about_us). The homepage contains a prominent announcement about the release of 2006 TRI data (searched May 2010), and users have access to TRI data under the heading, "Search for toxic releases," by inputting the name of the desired organization, the city where released, or zip code. The most general search produces TRI data, but in a different format from the EPA websites. Instead of a single chart, a mixture of graphs and tables present the following information: a summary, a series of "Top 5" tables (cities for pounds of onsite releases; parent companies for pounds of releases; chemicals for pounds of releases; chemicals for risk score (RSEI); general and specific industries for releases); pie charts

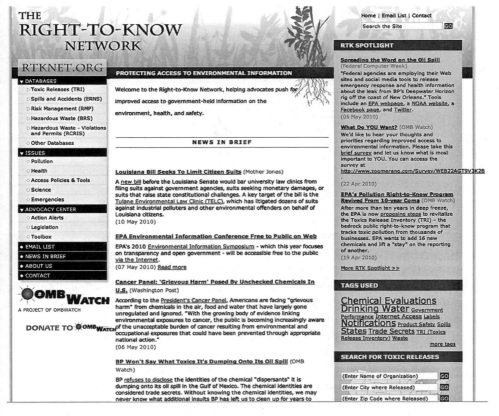

Figure 3.3
RTKNET.org homepage screenshot

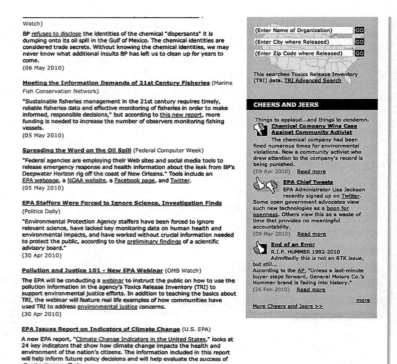

Figure 3.3
(continued)

describing waste released and generated; and bar graphs showing release trends per annum going back as far as 1997 (see image 3.4). The user can expand each of the charts and tables to include all entries, and comprehensive lists of facilities and submissions are hyperlinked in the summary. Lastly, the display can be changed globally according to level of detail.

An alternative method for searching the TRI occurs when one clicks "Toxic Release Inventory" under the "databases" tab. The website allows for an advanced search either by Facilities (options include "Facility and parent company," "Location," and "Industry and chemical") or by Waste Transfer (options include "Destination" and "Transfer Type") (http://www.rtknet.org/db/tri/search). Each of these subheadings prompts the user to enter appropriate search criteria. As promised on the homepage, RTK NET also allows the user to search other databases based on different criteria than toxic releases, including "Spills and Accidents (ERNS)," "Risk Management Data (RMP)," "Hazardous Wastes (BRS)," and "Hazardous Wastes (RCRIS)" (http://www.rtknet. org/db). All searches lead to the same presentation style of results. The added databases provide information on toxic spills and accidents (under "Emergency Response Notification System requirements"), risk management plans filed by relevant industries, the generation and shipment of waste, and hazardous waste permits (searched May 2010).

Scorecard (Green Media Toolshed)
In 1996, Environmental Defense staff pursued the task of putting the TRI data online (phone interview with Bill Pease, May 23, 2003). After its launch on April 22, 1998, Scorecard (www.scorecard.org) was

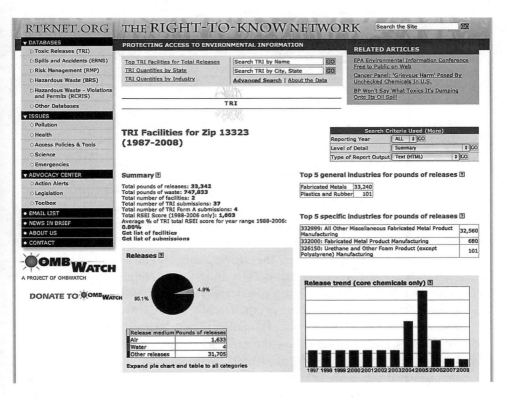

Figure 3.4
RTKNET report screenshot

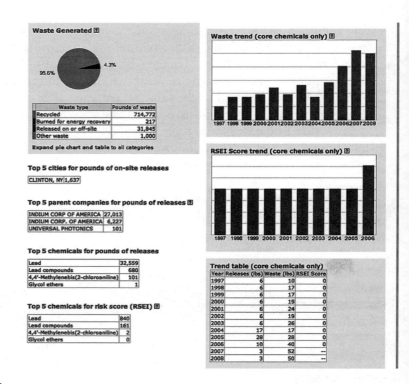

Figure 3.4
(continued)

expanded and refined (see image 3.5).[5] In 2005, Environmental Defense gave the website to Green Media Toolshed, an organization "committed to providing tools and improving the effectiveness of communications among environmental groups and the public" (http://www.greenmedia-toolshed.org/about/mission).

Scorecard's homepage stages the engagement as "The Pollution Information Site." Users are invited to "Get an in-depth pollution report for your county, covering air, water, chemicals, and more." Visitors can also investigate those topics in more depth, access "More Facts on Pollution," and take action in terms of EPA policy. Upon personalizing—being invited to enter one's zip code—the user views a page that offers menus in various categories: toxics, air, water, environmental justice, and "do more" (see image 3.6). The site visitor can choose among those headings, broken down into primarily comparative subheadings. For example, a user could follow the "See how your county stacks up" link, or see lists

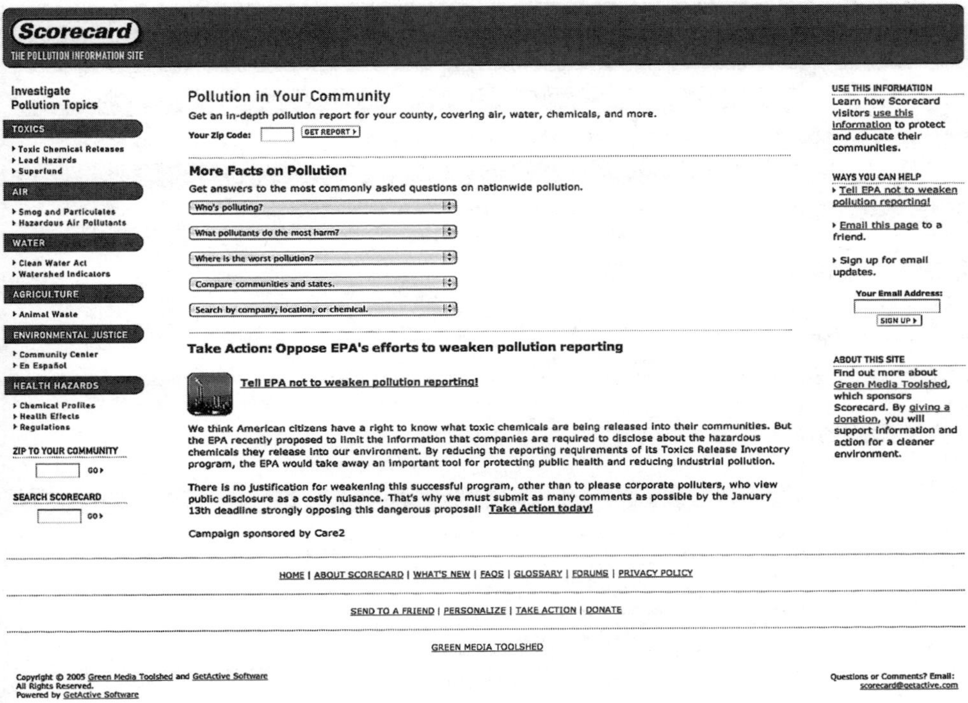

Figure 3.5
Scorecard.org homepage screenshot

of top polluters or top chemicals released in the county. To follow the list of top polluters, one finds a list of companies, ranked top to bottom in pounds of chemicals emitted. Click on a company name, and the user is taken to another list, this one offering maps, rankings of pollution in terms of potential health risks, a catalog of health effects, summaries of the data, a button urging people to take action, and links to other sites (searched May 2010).

Virtual History

The historical understanding of the problem of toxic pollution embedded within these websites continues the narrative of environmental justice as a problem of information poverty, following from the specific "lessons" of industrial hygiene and the community catastrophes of Love Canal, Woburn, and Bhopal. This digitized history can be seen both explicitly

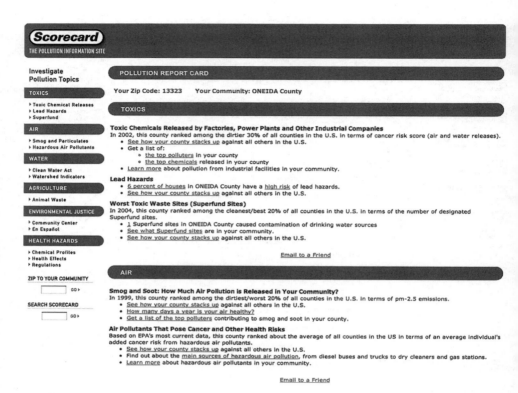

Figure 3.6
Scorecard.org report card screenshot

(referencing tragedies and political interventions) and implicitly (organizing the interface to solve the problem of an information deficit for the user).

The explicit history exists directly in the EPA's presentation of the Toxics Release Inventory—the database on which all these websites are at least partially based. The EPA characterizes the historical emergence of the TRI in the following manner:

In 1984, a deadly cloud of methyl isocyanate killed thousands of people in Bhopal, India. Shortly thereafter, there was a serious chemical release at a sister plant in West Virginia. These incidents underscored demands by industrial workers and communities in several states for information on hazardous materials. Public interest and environmental organizations around the country accelerated demands for information on toxic chemicals being released "beyond the fence line"—outside of the facility. Against this background, the Emergency Planning and Community Right-to-Know Act (EPCRA) was enacted in 1986. (http://www.epa.gov/tri/triprogram/whatis.htm)

As presented on the EPA's website, the demands by workers and community members leveled at the government in response to these disasters were not calls for less pollution, but for more information to manage risk. The legislation that followed those events, and led to the creation of the TRI, thus hinged on the community's right to know about waste disposal and emergency planning (Hadden 1994). In acting as database distribution centers, these websites became online incarnations of the specific trajectory of environmental antitoxics activism, organizing digital engagement by enabling connection to much-needed information.

The implicit history is found in the modes of intervention offered and the positioning of the site user as someone primarily in need of information. As with Alice Hamilton's work on industrial hygiene, the main villain is ignorance, not willful negligence or deceit.

Audience Composition: The Digital Divide and the Parameters of Place

Apart from building on histories of the toxics and environmental justice movements, the TRI websites compose audiences on the World Wide Web. In this manner, they also seek to craft new, virtual spaces where scientific information and community needs come together. In particular, web access serves as a constraint over the recruitment and retention of potential audiences. Recent data suggest that while access to the Internet in the United States has expanded, people of color and of lower income lag behind in terms of technology adoption such as broadband Internet at home.[6] Ironically for the mission of the TRI websites, the people who tend to lack access to computer technology (and thus websites) or those whose common usage patterns are less defined are the very same people most likely exposed to toxic pollution (see Bullard 2000). These are roughly the same groups that for years have lived on the "wrong side" of the digital divide (see, for example, Norris 2001). Related questions emerge regarding whether barriers related to web access—cultural patterns of information seeking, social networks that eschew online communication—could play a similarly discouraging role even if the entire United States were "wired."

For those able (and perhaps predisposed) to access the digital environment, each TRI website reaches discrete audiences based on geography. Specifically, users enter their zip code, county, or city/state in order to access a particular performance of the TRI—one's location is the price of admission. Even if this construction is strongly determined by the methodologies of data collection and categorization, other hypothetical boundaries would have very different consequences for audience construction. Three examples illustrate the point: (1) data organized by community (strongly locally determined boundaries) would reinforce the responsibility and power of communities to act; (2) data organized by legislative districts would emphasize the political component of approving, blocking, or remediating toxic production and transport; and (3) data organized by watershed or airshed would remind an audience of the ecological realities of pollution and a community's connection to ecosystems. Each of these hypothetical alternatives suggests different relevant histories about toxics. Data organized by community might emphasize the history of prior activism and engagement with corporate and governmental representatives. Data organized by legislative district might spur historical attention to the corrupt political machine that allowed so many toxic dumps to be sited in a particular area. Data organized with an eco-logic might motivate a focus on historical flows of pollution through an aquifer. These differences have consequences not only for what information is sought but for the kinds of constituencies that will tend to be attracted and thereby compose an audience (e.g., neighborhood associations versus environmental activists).

The impact of the digital divide and the power of defining the parameters of place serve as reminders that web-based knowledge is not simply "public" knowledge. Parts of the public are recruited and excluded in various ways that affect performances of the TRI.

Webs of Competency

In the virtual space of the websites, issues of audience composition strongly correlate with presumed audience competency. What skills do website visitors need to navigate the TRI? We find that the sites simultaneously emphasize their ease of use on the one hand and, on the other, the need to assess the limitations of the TRI data. This tension not only runs through the expectations of the website producers but through the websites' overtures toward crafting a new space of participation where citizens and data can come together.

The ease of the interface becomes an important selling point for each site, minimizing the presumed competencies (and time and dedication) required of the prospective audience. The first competency is simple: know one's zip code, the primary datum that welcomes the user to the site. Scorecard is especially replete with references to the ease of use: "Our goal is to make the local environment as easy to check on as the local weather. You may not do it all the time, but it should always be at your fingertips" (Krupp 1999). Importantly, other competencies have been lessened in comparison to non-Internet-based information retrieval; these websites help to mitigate the geographic space and staffing times that had limited access prior to this technological intervention—traveling to files physically located in Washington, D.C., within operative business hours. Scorecard is sympathetic: "We know how hard it is to get public information on pollution in your community, so Scorecard is set up to let you *find the facts about local pollution*, and what hazards that pollution presents, with just a few mouse clicks" (http://www.scorecard.org/about/txt/sc_guide.pdf; emphasis in original). The digitization and web interface allow data to be accessed *virtually* anywhere and at any time. As a result, the potential audience expands to include those competent within the virtual realm.

Just as the expected level of competency created by the zip-code interface expands the audience, potential gaps in the information actually available may in turn shrink it. TRI provides very limited data, anchored to basic regulatory requirements. The reporting requirements are restricted to certain chemicals, industries of certain sizes, and self-reporting contingencies. In addition, the database includes a consistent two- to three-year delay in publishing reports (true for all sites investigated here). Finally, there is little certainty about how one should interpret the information in terms of personal exposure and risk. The EPA's site explains the limitations this way:

Users of TRI information should be aware that TRI data do not reveal whether or to what degree the public is exposed to listed chemicals. TRI data, in conjunction with other information, can be used as a starting point in evaluating exposures and risks. The determination of potential risk to human health and/or the environment depends upon many factors, including the toxicity of the chemical, the fate of the chemical in the environment, and the amount and duration of human or other exposure to the chemical. (http://www.epa.gov/tri/triprogram/FactorsToConPDF.pdf)

The EPA's data suggest many different things, none of which effectively communicate the risk posed to members of the lay public. The individual user, on visiting the site, can learn about the estimated pounds of pollutants that are being emitted in their general proximity, but must correlate that data with distribution patterns, exposure rates, and actual releases (all very local and particular information) and also gauge the relative toxicity of each of those contaminants. Moreover, the numbers themselves—pounds emitted—do not reflect the comparative potency of the chemicals, nor do they provide an adequate sense of the dangers that may be posed.

Knowledge about individual exposure and about scientific understanding is missing from the EPA's site, knowledge vital for making the data mean something to the specific user. As such, the site exists as a small node in a vast network of knowledge and understanding, stretching from the very specific elements of individual experience to the universal reach of scientific declarations. Importantly, the data that has been presented via TRI, by itself, is presented to an audience that is presumed to be able to make the information meaningful. The meaning of those numbers must still be filtered on the ground (through local environmental conditions), in the body (through exposure rates), in the brain (through learning the various contingencies and regulations and science that govern the data's significance), and in the ether (through nested sites offering a continual regress of explanation). Either that, or the numbers provided may be targeted for other goals (see below).

Both RTK NET and (especially) Scorecard augment the inherently limited TRI data by including other sources of information related to toxic emissions, thus offering more potential data with each search.[7] Information-gap caveats remain, but the sites have slightly different approaches to the presumed competencies of the audiences they hope to attract. RTK NET identifies the deficiencies of TRI, while at the same time highlighting its benefits: "Facilities must report their releases of a toxic chemical to TRI if they fulfill four criteria. . . . Therefore, not all, or even most, pollution is reported in TRI. However, TRI does have certain advantages: . . . It is congressionally mandated to be publicly available, by electronic and other means, to everyone. This means that it's relatively easy to obtain TRI data and that the data is well-known" (http://www.rtknet.org/db/tri/about). How to navigate those deficiencies is something the users will have to be competent to do themselves or to seek out advice from the owners of the site: "We provide a user manual and other materials that help you access and search the databases (online access to these materials is available under help and documents). In addition, you are welcome to call us if you have problems accessing or searching the databases" (http://www.rtknet.org/about_us). RTK NET suggests that expertise is *available* for those that need it, through channels outside of the web itself. For another example, RTK NET offers a link that documents known uses of Right-to-Know data, but that does not in itself provide prescribed expertise. The strategy invoked by this site introduces competency activated outside the digital network—people can become engaged not just by entering a zip code and staying on the site, but also by stepping offline and communicating as a particular community member working to apply these data to a specific problem.

Comparatively, Scorecard works to provide much more immediate information within the website itself, especially with regard to health impacts of specific chemicals and potential health risks. Unlike the other two, Scorecard offers broader comparative rankings, not just in terms of a company's performance relative to other companies (as TRI Explorer and RTK NET allow), but in terms of a community's level of toxicity in relation to other communities. These comparative measures help to provide some easily identifiable interpretations of meaning.

In addition, Scorecard assumes that the user wants to learn—the website's architecture goes beyond data presentation, offering tools for understanding the specific importance of the data. Users who peruse the pollution locator function can access "risk assessment values" ("RAVs") to look up potential health effects and risks involved with various "recognized" or "suspected" toxicants.[8] This risk assessment framework translates various toxins into comparable equivalencies by creating Toxic Equivalency Potentials (TEPs) using benzene as a baseline for carcinogens and toluene for noncarcinogens.[9] This framework represents a controversial attempt to present *some* level of risk analysis in the context of these data. Industry experts contend that this comparative measure does not adequately reflect the risk involved, and may cause unjustified panic (Foster, Fairley, and Mullin 1998).

While Scorecard provides a more comprehensive approach, limitations still apply: "Scorecard does not cover all major environmental problems, sources of pollution, or potential exposures to toxic chemicals. Our profiles are limited to environmental issues tracked by authoritative national data sources. . . . There are many gaps in the coverage of national regulatory programs, leading to gaps in our understanding of local environmental problems" (http://www.scorecard.org/about/txt/caveats.html). This lack of completeness or comprehensiveness in the data in all the sites (even those that aggregate more information and make the service more robust) require that audiences make decisions about what these numbers mean.

The caveats offered by all the sites suggest that these sites do not presume that their audiences will have the competence to fill in the missing pieces themselves. Scorecard in particular emphasizes the scientific veracity and legitimacy of their information, meaning to provide online access to "credible scientific data" that is "scientifically bulletproof" and thus politically valuable as such (phone interview with Bill Pease, May 23, 2003). In the following section, we address what the audience—the community members of this virtual, scientifically informed space—is expected to be able to do in the context of these limitations of the provided data.

Audience Roles—Digital Citizens, Networks, and Activism

The crux of our analysis of the web as a new space for digitally enabled participation addresses these websites' ability to transform the possibilities for citizen participation. Thus, after environmentally conscious digital citizens acquaint themselves with their toxic communities through the lens of TRI data, the next step involves *acting* on that information. In constructing audiences of digital activists, the TRI websites offer (to greater and lesser degrees) means for putting the knowledge to work. Some suggested actions involve connecting to other groups and initiatives through networks of websites; others involve specifically targeted messages communicated to polluters, regulators, or government officials. Such prodding to take action reflects the attempt to influence the audience's role in the broad performance of EJ and toxic pollution.

All three TRI websites attempt to create forms of digital activism. The interfaces that they provide, which offer varying degrees of context for the information, create environments for the practice of antitoxics activism. While not showing a perfect linear relationship, our analysis below demonstrates how the three websites offer increasingly "activist" roles along a number of dimensions. These "invitations" to activism attempt to employ the web as a space of citizen participation, allowing users to acquire and then use information offered by agencies and scientists.

While TRI Explorer, as the website maintained by a governmental agency, constructs the weakest activist role for its audience, we should not trivialize the EPA's explicit faith in the empowerment that the database provides: "Armed with TRI data, communities have more power to hold companies accountable and make informed decisions about how toxic chemicals are to be managed. The data often spurs companies to focus on their chemical management practices since they are being measured and made public. In addition, the data serves as a rough indicator of environmental progress over time" (http://www.epa.gov/tri/triprogram/whatis.htm). Consonant with historical understandings of environmental health and toxicity, information is established as a *prelude to* empowerment. Data is a resource—arming communities—and also "spurs" corporate action, but without apparent agency. Overall, while describing a kind of engaged and activist outcome, the website offers few tools to transform information into action.

RTK NET positions itself more explicitly as a springboard for further activism, and not as a digital tool itself. The site promotes the data to an audience of established activists, or people already concerned with the state of their environment. The website declares, "With the information available on RTK NET, you can identify specific factories and their environmental effects; find permits issued under environmental statutes; and identify civil cases filed" (http://www.rtknet.org/about_us). Although the specific risks must be evaluated and worked through by individuals doing their own research, RTK NET does link to sites where such research could be accomplished. In particular, a user could attempt to correlate chemical names with their suspected health impacts by following the link to the New Jersey State Fact Sheets on hazardous substances (http://web.doh.state.nj.us/rtkhsfs/indexfs.aspx) or the EPA's Integrated Risk Information System (http://www.epa.gov/iris/). These correlations can suggest how dangerous, in principle, a chemical is, but do not translate neatly into risk. Finally, the website contains links to explicitly political analyses at their advocacy center, updated regularly (see http://www.rtknet.org/taxonomy/term/22). Importantly, the website places these links under the following heading:

Citizens have a right to participate in government decision making about public information access policies and strategies. Citizens also have a right to hold the government accountable for enforcing policies requiring public dissemination of information. This section of RTK NET provides citizens with additional tools with which to influence decisions made by the government regarding your health and safety. (http://www.rtknet.org/taxonomy/term/22)

Such documents and RTK NET's list of "Organizations working on RTK" inspire a critical and activist orientation to the TRI data.

In concert with RTK NET, Scorecard's digital environment enables users to do much more than just get numbers. The website implores users to

Find out about the pollution problems in your community and learn who is responsible. See which geographic areas and companies have the worst pollution records. Identify which racial/ethnic and income groups bear more than their share of environmental burdens. Then take action as an informed citizen—you can fax a polluting company, contact your elected representatives, or get involved in your community. (http://scorecard.org/about/about.tcl)

As Pease recalled in a phone interview (May 23, 2003), many of the people who phoned Environmental Defense did not want to be informed about where to find data on pollution, but rather to learn about how to use the information to good effect.

Scorecard presents a menu of diverse choices for learning and doing more. Several options encourage comparisons, which not only provide more meaning than simple numbers but also motivate an emotional response. After typing in a zip code for Madison, Wisconsin, the user learns immediately that "In 2002, this county ranked among the dirtiest/ worst 20% of all counties in the U.S. in terms of air releases" (http:// scorecard.org/community/index.tcl?zip_code=53703&set_community_ zipcode_cookie_p=t&x=0&y=0). Such information creates a sense of unfairness without demanding that the user learn extensively about particular exposure and toxicity data. Knowing that your zip code is worse than others constructs a frame for action to clean up your environment to some higher standard. The website also makes it easy to focus on the "top polluters" in one's county—a qualitative and accessible description that interprets the quantitative data for the user.

Scorecard differs from the other two TRI-based websites most in terms of prodding its audience to speak out in a political manner. The homepage, for example, includes a prominent link to a petition to the EPA under the heading, "Take Action: Oppose EPA's efforts to weaken pollution reporting."[10] The site also provides direct, dedicated mechanisms enabling individuals to engage in the political process—digital letter writing in the form of faxes and emails. These tools direct individuals through the site and into the world, through the lens of environmental antitoxics activism.

When the user clicks on the fax button, for example, text appears that will be sent to the particular company. The fax identifies the user, primarily, as a "neighbor potentially affected by emissions from your facility," and as a member of those "who may be exposed." This status of potential victim of the plant is presented as an enabling feature, granting the digital citizen some measure of authority to demand change or, failing that, at least more information. Thus, the user (and potential sender of the fax) is constructed as "Concerned Neighbor" in the sense of wanting to know more about what this company is dumping into the surrounding environment. Other communications that Scorecard offers follow similar patterns. The sender identifies as a local community member and, on the basis of information they have obtained about their community, agitates for change. Officials are urged to alter policies and speed up processes in order to make more information available and force companies to be more accountable for their pollution.

Early in its inception, Scorecard drew significant attention as a model for what the web could do (Graham 2000). Since that time, Scorecard itself has collected testimonies from grassroots organizations, members of the public, educators, and professionals in media and government, which tout the value and uniqueness of the Scorecard service (see http://www.scorecard.org/about/txt/survey_1.html). The site represents a move to web-based intervention techniques, creating a hub that not only distributes information, but attempts to create mechanisms for people to be empowered—by providing opportunities to learn and to share, and to give voice to their concerns about toxic chemicals.[11] In addition, by adding more context for the data (e.g., preliminary risk assessment, comparative rankings) and specific actions based on that information (e.g., faxes, emails, monetary contributions), the site channels people toward particular interventions. Thus, Scorecard is a technology truly meant to *digitally* empower and in some sense construct the online citizen activist through a performance of TRI data.

Discussion and Recommendations

Many of the contributions to this book speak to changes underway or made possible in scientific and technological practices at a broad level. Our cases illustrate the theme at a more precise level, addressing how EJ concerns have shaped specific TRI websites by carrying forward historical attention to information poverty and, in a methodological sense, by doing so in relation to an information-deficit model. We also understand the TRI websites as new spaces where citizens, scientists, and government agency personnel commingle. What our analysis shows, however, is that creating such spaces is but a part of a strategy to transform participation toward dialog and away from one-way communication inspired by the deficit model of the public understanding of science.[12] Our dramaturgical lens reveals how the provision of data and information varies to construct different types of audiences. Some of them are very much continuous with historical interventions that seek to rectify information deficits, in some cases with the potential to integrate data and action in exciting ways. We hope to push the architects of such interfaces beyond consideration of a "generic user" and toward sensitivity to the social configurations that impact and result from such interventions, enabling a transformation of how both laypersons and scientific and technical experts engage with one another. We also hope to enlarge the conversation in environmental justice by drawing attention to the relatively narrow digital instantiation of the information-poverty trope and, in another way, by pulling back the curtain on how such websites are staged.

Just as importantly, the analysis here seeks to offer practical suggestions for more effective web-based interventions surrounding TRI— encouraging a more ethically capable online space for citizen participation in the EJ movement. First, we need to tell different and more nuanced histories of the antitoxics and EJ movements. By remembering the constructions of prior audiences (factory workers), the exclusion of certain narratives (corporate negligence), and the political struggles to define the scope of the TRI, we can instead contextualize interventions such as TRI websites as political acts and not simply gateways to neutral data. This can help achieve the transformation of both the EJ movement and technological website design.

Second, we should acknowledge that the choice of medium involves some exclusion of possible audience members. While we certainly do not dismiss the value of online access to TRI data, we call attention to the costs of potential exclusion and offer two complementary options. In one way, TRI websites cannot be the sole vehicle for dissemination of data. It would be reasonable to staff a phone line at the EPA, for example, that would permit non–computer users to ask for help in retrieving data (that could then be faxed or mailed). In another way, if the Internet represents a fundamental public good as a source of democratic access, then (environmental) justice here must include real infrastructural improvements that allow for greater, if not universal, online access and educational initiatives that help people gain comfort with online research.

Third, TRI websites should note that zip codes pose aggregation oddities, and encourage audiences to reflect on more organic groupings (as a part of the "data-gap alerts" that the databases offer). Such awareness might provoke discussion to improve or expand data-collection methods that would allow aggregation more amenable to political or ecological intervention. GIS technology, for example, might permit more user-flexible manipulation of TRI data at some future point—allowing users to orient their data retrieval by community, legislative district, or ecoshed.

Fourth, TRI websites should make explicit that the gaps in knowledge are *not purely scientific* in nature. The TRI data are not sufficient by themselves, and competencies must be activated by encouraging people to participate in a dialog about acceptable risk, for instance, by contributing their particular, place-based experiences. Learning can go both ways—from actor (expert) to audience (layperson), and from audience to actor. A wiki-style component of TRI websites could enable a participatory audience not only to benefit from the expertise and research of other audiences, but also serve to build a virtual political movement that empowers otherwise dispersed and isolated users to make their own contributions. In fact, such an initiative exists in the advocacy realm at the level of organizing advocacy itself. "WikiAdvocacy" (http://wikiadvocacy.org), for example, offers guidance on how both to start and to manage an advocacy group with standard wiki access, which any website visitor may edit; an activist toolkit (http://activist-toolkit.wikispaces.com/) provides further details for such possibilities (see also Kezar and Lester 2009).

Fifth, and finally, we argue for placing information-poverty problems into more broadly conceived political activist contexts, moving beyond the information-deficit model. We see the potential for other kinds of roles to be assigned (or at least suggested) in the context of online TRI data. Tasks of accountability might be integrated beyond assumptions of public shaming, and users could participate in discussions of design, anticipatory regulation, or integrated monitoring. Expanding the vision of a space for engagement that moves well beyond an exclusive focus on the narrative of information poverty is central to the task of empowerment and therefore the goals of environmental justice.

Conclusion

This chapter engages the framework of audience construction to analyze the virtual provision of data from the Toxics Release Inventory. By situating these TRI websites within a history of environmental justice and applying a dramaturgical lens, the chapter makes two primary arguments. First, the historical dominance of the information-poverty narrative in environmental justice discourse deserves greater critical attention. As other studies in this book meaningfully point out, citizens need and value scientific data about toxicity and regulatory information about the production, transport, and release of toxic materials. However, the mixed success of these intervention attempts suggests that information provides an opportunity for more profound transformations, but is not sufficient in and of itself. Thus we argue that TRI data do not have agency. Instead, the audiences for such information represent the potential agents of change, and their empowerment is as critical to reducing the hazards of toxic pollution as the collection and dissemination of data.

Second, all interventions in the service of environmental justice, and especially the web-based interventions that form the subject of the current study, must take special care to consider how they construct their audiences. Such performances involve choices in presentation that impact the possible transformative nature of the performance on an audience, by tapping into histories, assuming competencies, including and excluding members of the public, and assigning roles for action. Analyzing these intentional and accidental choices both creates insight into the trajectory of the EJ movement and also provides strategic guidance to the architects of online activism. For scientists and other technical experts looking to aid the cause of environmental justice, the focus on audience and the analytical framework we expound should be seen as a *means* to be more reflective. Such reflection may include implicit assumptions that expand or curtail the number and kinds of people who attempt to use the information that scientists and engineers labor to produce.

We urge organizations that provide access to environmental science (e.g., TRI data) for the purpose of furthering the agenda of environmental justice to transform their public performances to become more inclusive, more sensitive to diverse competencies, and more explicit in joining knowledge with action. The variability in the three TRI websites we have analyzed in this chapter underscores the potential for transforming the role of nonexperts to engage meaningful scientific information in struggles against toxic pollution.

Notes

1. Gottlieb (1993, 170) has noted that community-based antitoxics groups have "a direct lineage to earlier urban and industrial movements." Also, while the EJ and antitoxics movements differ in notable ways *as* social movements, we treat them together in this chapter because they stand in the same functional relationship to the websites we analyze.

2. The community of Woburn discovered that its groundwater was contaminated by trichloroethylene and tetrachloroethylene, but such knowledge had to be generated by a local activist group (For a Cleaner Environment) rather than by the companies that dumped the chemicals, W. R. Grace and Beatrice Foods (see Brown and Mikkelsen 1990). The incident was later popularized by the film *A Civil Action* (1998), starring John Travolta.

3. If chemical names are trade secrets, and thereby protected from competitors, the EPA will not list the name of the chemical. In an interesting linguistic move, the chemical name becomes "sanitized" before release (see http://www.epa.gov/enviro/html/tris/column/sanitized_ind.html).

4. The 1989 date comes from the RTK NET site itself (see http://www.rtknet. org/about_us). Alexa.com, however, reports that the site has been active since 1993 (see http://www.alexa.com/data/details/?url=www.rtknet.org). We believe that the discrepancy reflects being online versus being on the web.

5. While Scorecard remains an available website, much suggests that it is no longer actively managed. First, the site appears to tap into TRI's 2002 data rather than updated information from 2006. Second, their "Take Action" link on their homepage takes the user to a "closed" petition that addressed a campaign ending in January 2005. Third, clicking on their fundraising link brings the message that "this fundraising campaign has expired" (searched October 2010).

6. http://www.pewinternet.org/pdfs/PIP_Mobile.Data.Access.pdf.

7. "Scorecard integrates over 400 scientific and governmental databases to generate its customized profiles of local environmental quality and toxic chemicals" (http://www.scorecard.org/about/txt/data.html).

8. Scorecard breaks down the toxins into recognized and suspected, relying on California's Proposition 65 (www.scorecard.org/health-effects/gen/hazid.html).

9. For carcinogens, TEP = [Added Cancer Risk/Unit Release of Chemical X]/ [Added Cancer Risk/Unit Release of Benzene]; for noncarcinogens, TEP = [Hazard Index/Unit Release of Chemical X]/[Hazard Index/Unit Release of Toluene] (http://www.scorecard.org/env-releases/def/tep_caltox.html).

10. We note, however, that the "Take Action" link is over three years out of date—the petition is "closed" and the website indicates that the EPA was "accepting public comments only until January 13, 2005" (http://www.thepetitionsite. com/takeaction/761256353?z00m=65884, accessed June 2008).

11. In a phone interview (May 23, 2003), Pease estimated that around a hundred thousand unique visitors came to Scorecard each month. Of those, he believed that only about 1 percent used the take-action functions. While this percentage may sound low, it represents a reasonable response rate for online activism and produced a thousand actions per month.

12. The concept of dialog, which emphasizes a symmetrical relationship between scientists and publics in terms of information exchange and learning, now guides the structuring of many events aimed at public engagement with science (for an overview, see Rowe and Frewer 2005; Lehr et al. 2007). Central to the idea of dialog are the value and necessity of including the perspectives of laypeople in complex, risk-based decisions (see Irwin 2004).

4

Experts, Ethics, and Environmental Justice: Communicating and Contesting Results from Personal Exposure Science

Rachel Morello-Frosch, Phil Brown, Julia Green Brody, Rebecca Gasior Altman, Ruthann A. Rudel, Ami Zota, and Carla Pérez

[Personal exposure science] develops strong advocates—people can speak for themselves. The data generated supports the claims, experiences, and demands that the community members bring to the podium in policy settings. Community members can say: "I know this because my home was tested and all these chemicals were found in my house!"
—Interview with environmental justice organizer involved in conducting personal exposure assessment study

It was a large intensive health study that left many of its participants feeling like guinea pigs. . . . No public presentation was made to the community regarding the results and only the participants with a serious medical condition received feedback. . . . From the community's point of view, the results of this study were not useful and it seemed to create more problems than it solved.
—Schell and Tarbell 1998

For nearly two decades, environmental justice advocates have pioneered innovative and strategic collaborations on scientific research to answer challenging environmental health questions. Very often the issues addressed are firmly situated in the realms of scientific uncertainty and contestation (Jasanoff 1987), as scientists and advocates generate and disseminate different forms of scientific data and expert knowledge aimed at (re)shaping environmental policy and regulation. The rapidly expanding field of personal exposure assessment and biomonitoring is an emerging epicenter of contested knowledge in environmental health. As environmental justice communities work with researchers to document the sources and pathways of chemical trespass in their homes and bodies, they are also faced with the paucity of health effects data for many of the pollutants studied, which raises ethical and scientific challenges for whether and how to report results to individual study participants. In the context of community-based participatory research focused on environmental justice questions, this means ensuring that exposure data are reported in ways that are meaningful and that elucidate potential paths for individual or collective action to protect health. In addition, the fields of personal exposure assessment and biomonitoring are themselves scientific research pursuits strongly shaped by the questions and advocacy of the environmental justice movement.

Exposure assessment has always been one of most methodologically challenging aspects of environmental health science and environmental epidemiology. However, in the last ten years, the field has made significant advances, particularly in the realms of biomonitoring and environmental sampling of various media, such as air, water, and food (Goldstein 2005). The efforts of environmental justice advocates to broaden the definition of environment to the milieus where communities live, work, and play have encouraged the exposure assessment discipline to widen its gaze from a previous focus on monitoring contaminants in the outdoor environment toward developing sophisticated and more accessible techniques for sampling myriad microenvironments where humans do that living, working, and playing. For example, exposure assessments for various compounds, including lead, pesticides, polycyclic aromatic hydrocarbons, and other toxicants attempt to characterize exposures to pollutants that are deposited from indoor and outdoor sources in and on objects found in homes, where humans spend a significant amount of their time. House dust, air, and soil sampling techniques are now widely used to assess exposures and characterize potential health risks from personal exposures that may have more than one route of entry into the body (e.g., inhalation, ingestion, and dermal) (Lioy, Freeman, and Millette 2002; Rudel et al. 2003).

Similarly, it is becoming increasingly common for advocates to use biomonitoring in community cohorts to study the effects of chronic toxics exposures. Biomonitoring techniques fall into three basic categories: identifying biomarkers of exposure, effect, and susceptibility (Metcalf and Orloff 2004; Goldstein 2005). Previously restricted to logistically challenging and costly academic research and occupational cohort studies, biomonitoring techniques, particularly for exposure, have become more practical, less expensive, and more widely available. Moreover, advances in molecular biology have made these techniques more sensitive, specific, and biologically relevant. This has resulted in the proliferation of biomonitoring studies not only among scientists in academia and in state and federal agencies, but also in environmental advocacy organizations and nonprofit research institutes that believe such studies can be helpful in promoting stronger environmental regulation.

In 1999, the Centers for Disease Control and Prevention (CDC) incorporated a nationwide biomonitoring program into the National Health and Nutrition Examination Survey (CDC 1999), which provided the first point of reference for average levels of chemical contamination in the United States. The CDC report, now updated every few years (CDC 1999, 2003, 2005, 2009), has triggered a renewed and widespread focus on biomonitoring as a method of examining chemical contamination in diverse populations.

However, the science of exposure assessment is not a panacea. Although media sampling can provide evidence of contamination of microenvironments, and biomonitoring is a direct indicator of human exposure to certain compounds and their metabolites, neither technique can generally be used to easily identify the primary source(s). As one biomonitoring study participant stated, "None of these chemicals come with a return address." Moreover, rarely are these techniques able to predict health outcomes. Indeed, the rapidly expanding capacity of sampling and biomonitoring techniques to detect chemical compounds at increasingly lower levels has outpaced the capacity of environmental health scientists to accurately interpret the meaning of these results for health and even subclinical effects in humans. Finally, as these techniques become less costly, researchers have expanded the type and number of chemicals being studied, yet many of these substances lack toxicological or epidemiological evidence regarding their potential health effects (Davis and Webster 2002) and may not have any regulatory benchmarks to which media sampling and biomonitoring results could be compared (Wagner 1997). This situation—in which the enhanced sophistication of exposure assessment techniques consistently outpaces scientific capacity to interpret the meaning of data results for public health—poses ethical quandaries for scientists, public health practitioners, and environmental justice advocates, concerning whether and how to report exposures when clinical effects are uncertain (Altman et al. 2008; Morello-Frosch et al. 2009).

In this chapter, we examine the evolution of exposure assessment science and the proliferation of biomonitoring techniques in the academic and advocacy arenas. We see that evolution as one outcome of the ruptures in environmental health conditions produced by the increase in microenvironmental contaminants over the past thirty years. In the chapter, we identify three approaches for how these methods and relevant ethical considerations have been appropriated by scientists, mainstream environmental organizations, and environmental justice activists to understand the impacts of low-level chemical exposures on human health and to leverage regulatory and policy change: (1) clinical ethics, a biomedically driven approach; (2) community-based participatory research (CBPR), a prevention research–focused approach; and (3) citizen science "data judo," an advocacy-driven approach. These approaches help illustrate the relationships between the values of the environmental justice movement and changing scientific practices. While the clinical ethics framework is more commonly used, the CBPR and "data judo" frameworks are emerging as more effective strategies that open opportunities for democratizing environmental health science research and advancing environmental justice in the realms of organizing and advocacy. Nevertheless, actors raise significant ethical and scientific challenges when considering whether and how to report individual exposure information when implications for health and exposure reduction are uncertain. We then report on what government publications, professional association "best practice" guidelines, and the scientific literature offer to research teams considering individual report-back, followed by recommendations gleaned from interviews with researchers currently collecting and reporting individual exposure data. Finally, we discuss some of the ethical implications for future work in this area, assessing opportunities and obstacles for better linking science with policy and regulatory change by drawing on experiences from our research/activist collaboration to discuss household environmental sampling and relevant report-back strategies. Our collaborative includes scientists and social scientists from Brown University and the University of California–Berkeley who study environmental health and justice issues; Silent Spring Institute, a community-based organization that researches women's health and the environment; and Communities for a Better Environment, an organization that conducts organizing, advocacy, and litigation related to environmental justice issues in the San Francisco Bay Area.

Background and Methods

Communities that are socially, economically, and politically marginalized— from Native American communities in Akwesasne, New York, and St. Lawrence Island, Alaska, to African-American communities in Anniston, Alabama, and New Orleans—are beginning to conduct biomonitoring and personal exposure research to record the extent of community-specific contamination, and leverage government funding, industry action, or legal remedies. However, environmental justice advocates have approached biomonitoring and similar household exposure studies with caution because of concerns that "after-the-fact" measurements cast communities as environmental hazard detectors (Bhatia et al. 2005). Furthermore, this strategy can potentially "overscientize" environmental health problems, overlooking upstream causes rooted in social inequality, economic exploitation, and racial discrimination (Sze and Prakash 2004; Morello-Frosch et al. 2006). These political contexts add to the existing challenges researchers and environmental justice advocates face regarding health effects and knowledge of exposure sources and pathways.

Increasingly, researchers and environmental justice advocates also are confronted by the ethical challenge of whether to notify participants of individual exposure results when this information may not offer insights on individual health conditions, or where the source and pathway of individual exposures is not easily elucidated. Our interest in these scientific and ethical report-back challenges stems from our own research collaborative that connects environmental justice organizing with research and advocacy on environmental linkages to breast cancer. Specifically, the study entails environmental sampling of household air and dust in two study areas (Cape Cod, Massachusetts, and Richmond, California), as well as biomonitoring in one of those sites (Cape Cod) to assess the presence of endocrine-disrupting chemicals and other pollutants potentially linked to breast cancer and respiratory health outcomes (Rudel et al. 2003; Brody et al. 2007). We have chosen to report aggregate exposure assessment results through peer-reviewed publications, media outreach, public meetings, and contacting study participants individually.

As we developed our research project, we wanted to assess what information on report-back protocols was available to researchers and communities in order to stimulate dialog around report-back issues and develop best-practice guidelines. This included examining exposure reports published by government agencies, "best-practice" guidelines issued by professional associations, and journal articles on individual studies. We also interviewed scientists and community members who carried out and/or openly participated in biomonitoring and personal exposure assessment research. We assessed how these documents presented exposure data, and what information, if any, was provided about interpreting and acting on the exposure data. We also examined how scientific uncertainty and data gaps were explained to communities and study participants. We interviewed other scientists and research teams doing exposure studies to see how they made decisions regarding report-back of biomonitoring information. We began by contacting colleagues involved in academic and advocacy biomonitoring research, and added to our sample additional researchers that our colleagues recommended, through a snowball sampling method. Our data come from twenty-six interviews (lasting between an hour and an hour and a half), a review of relevant literature, and participant observation at conferences and workshops where these report-back issues were debated and discussed. Uncited quotes and information come from those interviews and observations.

Surprisingly, we found little guidance or "best-practice" resources aimed at scientists and academic-community research collaboratives that want to report individual and community-level exposure data to study participants. Although public health professionals have developed methods for reporting to individuals on regulated contaminants, such as lead, study participants often are not informed of their personal results if they do not have regulatory or clinical significance (Schulte and Singal 1996). Apart from the disclosure of information on a few well-studied contaminants like lead, the issue of reporting individual-level data to participants has traditionally been more of a concern in clinical medicine. In addition, public health researchers are used to studies with community-level data, such as cancer registry information or environmental contamination data in media such as food and water, and due to their large size, these studies are not logistically conducive to reporting data back to individuals.

The first document to specifically spell out the need to provide participants with information derived from their data was the Belmont Report, published in 1979 by the Department of Health, Education and Welfare (National Institutes of Health 1979). Since the Belmont Report played a formative role in modern human-subject protection, we may view report-back as linked to that historical legacy. As the technology used to detect environmental exposure has become more advanced, new ethical dilemmas have emerged regarding the reporting of data for which exposure is well documented, but health outcomes related to that exposure are uncertain or not scientifically validated. As a powerful and scientifically contested method, elucidating the ethical and policy implications of personal exposure assessment and biomonitoring is critical for providing guidance to those academic-community collaboratives that design research protocols and that are also faced with the daunting task of interpreting uncertain data and making decisions about how to protect health.

Frameworks for Approaching Results Communication in EJ Exposure Assessment Studies

A consensus has yet to emerge regarding the ethics of reporting individual data on environmental exposures when the relationship between exposures and health outcomes is not established. Indeed, some activists and scientists who generally support a broad right-to-know for study participants remain wary of the implications of individual notification of data when the clinical implications of the information are uncertain. For example, the emergence of several recent studies on the presence of polybrominated diphenylethers (PBDEs), polychlorinated biphenyls (PCBs), and other toxins in breast milk has raised such issues in light of the many known benefits of breastfeeding (Arendt 2008). The literature and our interviews with scientists and study participants conducted for this study, although not unequivocal, indicate a trend in favor of addressing report-back strategies in the recruitment and consent process for research studies. Our interviews, observations, and assessments of the literature found three main frameworks for reporting back biomonitoring results: (1) clinical ethics; (2) community-based participatory research; and (3) citizen-science data judo. All of them grapple with core issues of data interpretation, scientific uncertainty, confidentiality, and the right-to-know; all of them have grown with attention to concerns for ethically tenable practices, including questions of justice and democratic participation in the production and exchange of information.

Clinical Ethics

Clinical ethics suggests that the decision about whether to report individual biomonitoring results should be based on whether the risk relationship between exposure and health effects is understood (Shalowitz and Miller 2005). For biomarker levels for which an exposure-health outcome relationship is known, the clinical action level, or "the level at which biomarker results will be of concern," should be determined prior to the start of the study (Deck and Kosatsky 1999). If the results fall below this clinical action level, individual data should not be reported to participants. The clinical medicine model gives more weight to the expert-researcher's role in avoiding possible harm to study participants from reporting uncertain information and less weight to the study participants' ability to process complex and uncertain scientific information and respond autonomously.

The clinical approach does not permit precautionary action by participants whose biomonitoring or household results may approach but still be below an "action level," or regulatory benchmark of concern, even if the evidence suggests that there are health effects below the action level, as in the case of lead or mercury. Moreover, the clinical ethics framework offers a narrow view of the potential for beneficial action— usually focused on medical intervention or public health interventions based on regulatory guidelines or a legal mandate (such as child lead screening). In certain instances, these regulatory benchmarks may themselves be legally or scientifically contested.

In practice, the clinical ethics framework can overlook the significant evolution of clinical communications, particularly since patients have become more proactive in directing their own health care, often by tracking screening results, such as blood pressure and cholesterol, even when levels fall below a clinical action criterion (Bury 2004; Shalowitz and Miller 2005). In addition, the potential for individual-level data to provide relevant information on an individual's health is further complicated by the possibility of future scientific advances in establishing links between exposure and health outcomes. Indeed, as one academic research scientist interviewed stated, "[Individual results] are part of their medical history, so potentially in a few years that might be useful information."

Because no health effects are conclusively linked to individual low-level exposure for the majority of chemicals tested in biomonitoring studies, this clinical framework would likely lead researchers to report data only on an aggregate level. Nevertheless, our interviews with three medical doctors conducting biomonitoring research suggest a potential evolution of the clinical ethics framework. Based on their experiences as practitioners and researchers, they realized the advantages of engaging participants openly about their biomonitoring results and the associated scientific uncertainties as part of good "doctoring" and productive clinical interactions.

As more information on average levels of population exposures becomes available, researchers will have more possibilities for interpreting individual data. While regulatory benchmarks are unavailable for most of the chemicals tested in humans, comparing individual results with information on national-average exposures in the general population provide one way for participants to understand the meaning of their individual biomonitoring results. Indeed, scientists involved in an epidemiological cohort study of the developmental effects of pesticide exposures explained that the research team began their investigation by only reporting aggregate biomonitoring results; however, the investigators subsequently changed course and decided to make individual-level results available to study participants because exposure levels could be meaningfully compared to the national-average levels provided by the CDC's biomonitoring report.

Community-Based Participatory Research (CBPR)

The CBPR approach assumes that individual- and aggregate-level reporting of study results can empower communities and individuals to act on scientific evidence (Bishop 1994). One of the primary aims of the CBPR approach is the restructuring of unequal and discriminatory power relationships in society (Wallerstein and Duran 2003). The approach stipulates that the sharing of knowledge (such as biomonitoring results) between researchers and participants can have an impact beyond the scientific relevance of the knowledge for individual health (Foucault and Gordon 1980). Therefore, CBPR encourages as much information dissemination as possible to study participants, and posits that ownership of collected data lies primarily with the participants from whose homes or bodies the original samples were taken (Bishop 1994). CBPR researchers understand that a balance must be reached between time and energy spent reporting results back to communities and reporting results in scientific forums (Sullivan et al. 2001; Israel et al. 2003).

Researchers reporting pesticide exposure results to farmworker families in North Carolina echo this approach, stating that "communicating risk to affected individuals should be an integral part of any community-based project. It is ethical to return information to the owner of that information" (Quandt et al. 2004, 642). Indeed, investigators in this farmworker study assumed that individual report-back for all chemicals analyzed would occur, and therefore the main question was not *whether* to report individual results, but *how*. According to this CBPR framework, even information about an exposure for which a corresponding risk relationship is not available can have some benefits to participants, such as taking action to reduce personal exposures. The North Carolina study emphasizes community involvement in the development of report-back protocols to address the interests and concerns of study participants: "In terms of the ambiguity, [the participants] thought it was important that the scientists present *la verdad* (the truth). If this meant telling women that it was not possible to know the level of danger represented by the findings, they would prefer to know that rather than to have the scientists give them a simpler but incomplete answer" (Quandt et al. 2004, 638).

Thus, the CBPR approach to reporting data assumes that results should be disseminated to participants not only to communicate health information, but also to address disparities in access to knowledge that traditionally characterize lay-professional relationships (Sullivan et al. 2001). In this sense, CBPR is attentive to promoting equal participation in the knowledge-collecting process and nondiscriminatory power relationships. The approach must be strategic, however, since this framework raises potential conflicts of community versus individual right-to-know. The broad dissemination of biomonitoring results can adversely affect communities under study, even if the rights and confidentiality of individual study participants are protected. For example, if a population subgroup is found to have a higher frequency of a genetic polymorphism that confers susceptibility to an environmental exposure, they may be collectively or individually stigmatized. Individually, they may be denied jobs or health or life insurance if they are associated with an "at-risk" population. Collectively, a community perceived as "at risk" or "contaminated" may be passed over for programs or benefits, may face stereotyping that affects the quality of health care, or may suffer lost real estate values or financial liability for remediation (Weijer 1999). For example, as early news broke of elevated PCB levels in the community of Broughton Island in northern Canada, and before the full extent of contamination was understood to extend throughout the circumpolar region, Broughton Islanders were initially shunned as the "PCB people" with an adverse impact on the livelihood of the fishing community (Colborn, Dumanoski, and Myers 1996, 108). The potential pitfalls of report-back can be elucidated and more effectively addressed if researchers purposefully seek to develop study protocols and communication strategies in partnership with communities that are participating in biomonitoring studies (Cone 2005; Brody et al. 2007).

Citizen-Science "Data Judo"

We use the term *data judo* or *advocacy biomonitoring* to refer to a strategy in which study design and individual-results communication are shaped primarily by policy goals to improve chemical regulation (also see Morello-Frosch et al. 2009). Indeed this framework assumes that personalized information about chemical body burden can broaden public support for toxics use-reduction policies, and motivate individuals to engage in collective activism and also pursue individual exposure reduction. Environmental advocacy groups and communities marshal their own scientific resources and expertise to conduct research, and report-back strategies are specifically aimed to advance regulatory and policy change (Morello-Frosch et al. 2005). Interviews we conducted with scientists who designed and carried out body-burden studies for environmental organizations, as well as the participants in these studies, support this framework.

Although the data-judo approach to report-back has overlapping goals with the CBPR framework, there are some important differences. While CBPR primarily aims to use report-back strategies in order to break down power and knowledge disparities between scientists and communities, the data-judo approach explicitly seeks to mobilize constituencies by increasing public awareness around a specific regulatory issue or policy initiative. Over the past five years, there has been a proliferation of body-burden studies spearheaded by environmental organizations. Three milestone activist body-burden studies were conducted by the Environmental Working Group (EWG). The first, known as the Body Burden Study, recruited nine volunteers, most of whom were prominent environmental advocates, to have their blood and urine tested for the presence of 210 chemicals commonly found in consumer products and industrial pollution streams (Environmental Working Group 2003a). An average of 91 industrial compounds, pollutants, and other chemicals were found in the blood and/or urine of the study participants, with a total of 167 chemicals found in the entire group. The report on this study appears on the EWG website, where viewers can click on a thumbnail photo of each study participant to see what contaminants are in that person's body. The release of this study raised public awareness about the ubiquity of chemical exposures from diverse sources and the inadequacies of current chemical regulation regimes.

The second EWG study examined the presence of a category of brominated flame retardants (PBDEs) in the breast milk of twenty first-time U.S. mothers (Environmental Working Group 2003b). This study found an average level of bromine-based chemicals in breast milk that was seventy-five times the average found in recent European studies (Norén and Meironyté 2000; Strandman, Koistinen, and Vartiainen 2000). Milk from two study participants contained the highest levels of fire retardants ever reported in the United States, and milk from several of the mothers in EWG's study had among the highest levels of these chemicals yet detected worldwide. The third study examined umbilical cord blood, finding 287 different chemicals in 10 babies born between August and

September 2004 in U.S. hospitals (averaging 200 pollutants per infant). Of the 287 chemicals, many of which are found in consumer products such as stain repellants, fast-food packaging, clothes, and textiles, 180 were known carcinogens, 217 were toxic to the brain and nervous system, and 208 caused birth defects and abnormal development in animal tests (Environmental Working Group 2005).

The emergence of advocacy biomonitoring has made the image of ubiquitously contaminated bodies resonate widely in the media, regulatory, and policy arenas, and has led to a proliferation of studies by other environmental organizations and media outlets, including Commonweal, the World Wildlife Federation, Greenpeace, Environmental Defence (Canada), the Sightline Institute, National Geographic, CNN, and a major newspaper in Oakland, California (Sightline Institute 2004; World Wildlife Federation–UK 2004; Environmental Defence 2005; Fischer 2005; Greenpeace International 2005; Body Burden Work Group and Commonweal Biomonitoring Resource Center 2007; Duncan 2006; CNN 2007).

Unlike the other two frameworks, advocacy biomonitoring studies collect personal exposure data explicitly to highlight the failure of environmental regulations and policies, such as the Toxics Substances Control Act (TSCA), to protect the public from exposures to ubiquitous contaminants, most of which have not been tested to assess their potential short- and long-term health impacts. Many of these studies also find pollutant levels in study participants that are high enough to warrant health concerns, raising questions about whether current regulations are effective at protecting public health. As a result of extensive public outreach by both organization scientists and study participants, advocacy biomonitoring has garnered extensive regulatory attention, and legitimated mounting public concern about the ubiquitous presence of these chemicals in consumer products and their movement across diverse environments (Iles 2007). With few exceptions, these advocacy studies report data to study participants individually and also provide opportunities for them to talk publicly about their results. For example, EWG provides online personal biographies of study participants in their Body Burden and Breast Milk studies (Environmental Working Group 2003a; Environmental Working Group 2005). Many of these biographies emphasize that participants have made efforts to lead "healthy" lifestyles and that, like most of us, they did not work directly with chemicals on the job or live near major pollution sources. Participants in advocacy biomonitoring studies savored the opportunity to share their results with other study participants to better contextualize their meaning and highlight opportunities for exposure reduction. As one participant noted,

The important thing, I think, to me, was understanding my results in the context of other people's results. So that while each of us got our results individually . . . it was only sort of when most of us [study participants] agreed to be in a conference call together to talk about it that I sort of began to understand what my own results meant, and how I felt about it in the context of other people's reactions. . . . And so it was very important to me that as a group we agreed to share our results. Not that we now know exactly what it means, but it was interesting to note that the biggest fish eaters had the highest levels of mercury.

There are controversial aspects of advocacy biomonitoring compared to more traditional forms of biomonitoring report-back protocols. One of the more controversial aspects is that it explicitly challenges traditional Institutional Review Board (IRB) protocols of protecting participant confidentiality by giving study participants opportunities to discuss their results publicly, with the media and with each other. Based on our interviews with academic scientists, many IRBs have traditionally allowed aggregate reporting of study results, while restricting or strongly discouraging the conveyance of individual-level information. For example, some academic IRBs require passive individual report-back protocols, which prohibit researchers from proactively contacting participants to ask if they want to receive results. Although IRB concern about participant confidentiality is warranted, report-back protocols that require greater initiative on the part of study participants to acquire their results ignore the fact that many individuals want their own data in order to take individual or collective action to reduce exposures. Participants may also want to share their personal results with other study participants or collectively through their own networks, communities, and public forums. As one scientist from an advocacy organization argued, "I think part of the challenge for all of the biomonitoring studies that are going on, including ours, is that you want to do it by the book, so that you write up an IRB [protocol] like any other study with human subjects, but in a way, doing it by the book is exactly what this is not about." Therefore, some advocacy biomonitoring studies have encouraged IRBs to examine how traditional standards of confidentiality may impose problematic restrictions on individual report-back protocols. The studies in this sense call into question the ethical commitments that follow from, in the scientist's words, "doing it by the book." For some communities, these restrictions can be perceived as undermining the capacity of study participants to understand the implications of the study and to take protective action by first comparing their individual results in the context of those of their peers.

Central Issues in Reporting Exposure Data to Individuals and Communities

The above discussion of these three approaches to report-back leads us to formulate some general guidelines for reporting exposure data to study participants. Apart from distinctions between the three, all researchers, government agencies, and activists face common concerns in their work on biomonitoring and personal exposure studies. By taking these concerns seriously, we identify six key lessons that can be taken away by all involved parties and leveraged in ways that impart a strong environmental justice emphasis in this growing area of work.

The first of these lessons concerns setting expectations for what studies can say and how participants can use their results. Our research suggests that the ethical issues of reporting back exposure monitoring results necessitate addressing the rights of study participants to information before, during, and after studies so that they can make informed decisions and take action. Study participants often want their individual results and an interpretation of them in terms of what potential exposures may mean for their health. However, researchers and public health practitioners face ethical issues in interpreting exposure results when health and safety data are not available for the pollutants under study or when there is no scientific consensus about the risks associated with exposures. Perhaps the most important issue to emerge from our interviews with academics, scientists from advocacy organizations, and personal exposure study participants, some of whom were involved in EJ-focused projects, is that it is desirable to set expectations for any exposure assessment or biomonitoring study *before* commencing data collection and setting up report-back protocols. One important aspect of this effort is to clarify the inherent scientific limitations of interpreting what the data collected could mean for community and individual-level health. Equally important, even if health implications are unknown, individual-level report-back can provide an impetus for people to take individual action that could reduce their exposures. It can also provide participants with opportunities to collectively leverage results to support advocacy that promotes broader biomonitoring efforts to fully understand population variability in exposures, or interventions that promote tougher regulation or toxics use-reduction strategies.

A second area for change is to provide background information to make individual results meaningful. Several scientists and participants liked comparing individual data with aggregate study results. Such comparisons seem important for placing the information into a familiar context, as Quandt et al. (2004, 642) found: "Presenting individual exposure data with reference to actual community data, rather than more abstract population-level reference data, engages community members' interest." The use of comparisons is also recognized in the literature on risk communication as particularly important when the values being communicated appear small, or when risks are unfamiliar to the community involved in the study (Williams 2004). Body-burden studies can fit both of these criteria: chemicals are often detected in seemingly low concentrations, and they may involve chemicals unfamiliar to the general population.

A similar system for reporting individual-level data is to compare it with other published studies, such as the CDC reports (CDC 1999, 2003, 2005, 2009). It is important to keep in mind, however, that there can be some confusion about what this average comparison data implies. For example, one researcher indicated that when pesticide exposure results were reported to individuals, it was critical to ensure that any comparisons to average population levels from the CDC report were not misinterpreted as safety benchmarks. In this way, the mean exposure for the population often acts as a substitute "population norm" (Deck and Kosatsky 1999; Quandt et al. 2004). This can have two potential negative effects on the participants' understanding of their risk: (1) it can lead to a false sense of security, with participants who have exposure levels higher than the general population comparing themselves to others with extremely high levels, and (2) it can lead to unnecessary concern when those with higher exposure levels than the study average assume that they have unsafe levels, regardless of the fact that the entire cohort might have levels significantly below levels that indicate cause for concern. One scientist we interviewed, who directed an exposure study on brominated fire retardants, indicated that two study participants had extremely high levels of PBDEs in their tissue samples. This caused at least one participant to be concerned, despite the fact that there was no indication that her results posed health risks for her or her child. The researcher described her reaction:

The participant who had the second highest result was really pretty blown away by it. She had done the study expecting that she would be one of the more healthy, safe, you know, protected. . . . It's really an unfortunate part about enrolling [participants] in studies and giving them results about contaminant levels in their bodies when you don't have an even distribution or a way that would kind of predict or prepare them for where they might be in that distribution and she took it really hard . . . the rest of the . . . [participants] felt lucky and felt protected.

Using study or population exposure averages as a means to interpret individual-level data clearly has potential pitfalls. However, this fact should not prevent group averages from being reported in the context of individual-level results. Whenever possible, information about group averages should be coupled with an explanation of what is known about the average levels in the general population, or information about relevant regulatory benchmarks.

A third take-away from our study is that we need to balance individual versus aggregate communication of results. Our interviews revealed that the process for developing report-back protocols varies widely, both among academic and advocacy biomonitoring studies. Some researchers develop report-back protocols with little community input, while others solicit significant input from the study community, scientific colleagues not directly involved in the study, and social scientists. Most interviewees acknowledged the importance of having community representatives involved in the decision of whether and how to report individual and aggregate study results. They felt that it should be the community's decision whether individuals receive their own data, especially in situations when studies included participants whose illness was potentially linked to a substance under study. Nevertheless, academic studies or research involving community-academic collaborations must adhere to the law that all entities that receive federal funding for research must operate in accordance with federally prescribed IRB procedures; this makes IRBs the final arbiters of whether or not to approve individual-level notification of study participants about biomonitoring results. The academic scientists we interviewed reported a wide variation in the willingness of their IRBs to grapple with the bureaucratic and logistical challenges of reviewing and approving individual-level report-back protocols for biomonitoring studies. In addition, some of the scientists we interviewed described a lack of consensus among study collaborators, including academic scientists and members of community advisory panels, about whether to report individual data. The disagreements over how to design report-back protocols show that, even when a commitment to right-to-know and community-based research exists, choosing which information to report to individuals in biomonitoring and exposure assessment studies may not be simple to negotiate among collaborators. For example, physicians sitting on an advisory board for one biomonitoring study tended to discourage individual report-back due to concerns that patients may have health-related questions linked to their study results that most doctors could not realistically answer. Conversely, community advocates and some industry representatives tended to favor releasing individual results to study participants, viewing this as a right-to-know issue.

Our fourth lesson speaks to choosing appropriate risk-communication strategies. Most scientists described a system of individual and aggregate report-back that involved a combination of written materials and conversations with experts, either over the phone or in person. Some had a form of passive reporting, where study participants could contact researchers if they wanted to confidentially receive their personal results. This system also gives participants the opportunity to opt out of receiving their individual-level information. Another researcher stressed the need to follow up report-back with support from a counselor and/or to have someone contact participants down the line when questions arise related to emerging health issues or new concerns. One scientist discussed the need to remain extremely flexible and available for participants, since a third of the participants who did not opt to call in for results later expressed interest in getting their results, during a follow-up survey a few weeks later. This demonstrates that passive reporting is imperfect in providing results to participants who want them.

The report-back process offers the potential to use aggregate and individual-level information to develop exposure-reduction interventions. Indeed, receiving information about how to remove pesticides from the home or how to prevent future contamination was reported to be the most important part of the report-back process in two pesticide exposure studies. The promotion of public health interventions that are directly related to study results is an opportunity for scientists to ensure that the information provided to participants has a positive effect on their ability to take action to promote their health and well-being. One scientist brought up the importance of reporting individual study results in combination with specific exposure-reduction recommendations that participants can follow individually: "The most important component of that for us was not only giving the information but giving information about what the women could do. So that reporting back is always linked to action, so that they are not getting the information without having any idea of what they can do about it."

In one pesticide study, the health workers explained direct actions that all women could take to prevent pesticides from entering their homes and getting picked up by children, including closing windows during crop spraying and having farmworkers change clothes before entering the home. In addition, brochures were provided, with information in Spanish, about storing and washing work clothes separately and the idea of pesticide residues being invisible (Deck and Kosatsky 1999; Quandt et al. 2004). Other biomonitoring studies of persistent organic pollutants that bioaccumulate up the food chain provide participants with information about how to reduce their consumption of animal products or decrease the presence of contaminants in household dust by switching to less toxic consumer products. However, scientists are often forced to balance the potential disruption and cost of an intervention with the strength of the information indicating a pollutant's origins and health impacts. For example, one scientist leading a study on brominated flame retardants indicated that he would provide participants with information on how to reduce levels of animal fat in their diet, citing other health benefits associated with this action. On the other hand, he also indicated that he would refrain from advising participants to take more costly or inconvenient action to minimize the presence of PBDEs in household dust. This is especially true if the effectiveness of these interventions has not been assessed. As the scientist observed, "Right now my gut feeling would be not to tell people you should throw away all your furniture and buy all new furniture. That seems kind of extreme, right?"

Thus, it seems that in the case of PBDEs, for which the health effects are not thoroughly understood, the decision of whether to provide health information is related to the disruption that the health intervention would cause, and whether the intervention has other public health benefits besides minimizing pollutant exposures, such as reducing animal fat content in the diet, which can reduce the risk of heart disease.

There are certain instances when biomonitoring results raise conflicts with an existing public health practice with a known health benefit, as with breastfeeding. While there are indications that PBDEs may pose potential developmental health effects on offspring, toxicological evidence suggests that most of these effects occur in utero rather than through exposures through breastfeeding (Hooper and Sze 2003). However, breast milk studies have been controversial because of concerns that they may discourage breastfeeding, despite its known health benefits. To our knowledge, whether biomonitoring and public concern about chemicals in breast milk have changed breastfeeding behaviors has not been documented. Nevertheless, this issue has created controversy among public health advocates. To respond to this debate, a recent article proposed a model informed-consent protocol for breast milk biomonitoring studies that includes "advice that breastfeeding is almost always considered to be the best form of nutrition for a baby, and that the fact that the study is being carried out should in no way be taken as implying anything to the contrary" (Bates et al. 2002). All three scientists we interviewed who were involved in breast milk studies reported that they encouraged participants to breastfeed. Empirical investigation of mothers' responses in breast milk studies that encourage breastfeeding could inform the design of future monitoring efforts.

Finally, debates over "risk messaging" related to biomonitoring research are nowhere more difficult than in cases where health implications warrant exposure reduction, but interventions are either impossible or unjust, or would produce other, perhaps more deleterious consequences. In the 1980s and early 1990s, communication of biomonitoring results among Arctic Inuit communities called into question the consumption of their traditional food source of large marine mammals. Contaminants bioaccumulate and are delivered through many marine mammal food sources that are essential to community survival, subsistence, and the viability of a hunting culture. In this context, the conundrum lies in the paucity of viable alternative foods sources. Imported, market-based foods pose their own, arguably more dire health consequences in the form of malnutrition, obesity, cardiovascular disease, and diabetes (Frugal et al. 2003; Cone 2005). Increasingly, messages tend to encourage consumption of particular species with lower contaminant levels or specific cuts of flesh, a strategy that resonates with the case of Latino, African-American, and Hmong populations and fish contamination in Madison, Wisconsin, discussed by Powell and Powell (chapter 6, this volume). Yet, mounting evidence of the reproductive, immunological, and developmental effects of these persistent contaminants leaves many communities and scientists in an uncertain situation in terms of report-back strategies (Cone 2005). Scientists and community members involved in these studies support community right-to-know. But this work also poses significant challenges because of the environmental persistence of some contaminants, with solutions potentially a long way off or unachievable, and because exposure-reduction strategies can raise new problematic consequences (Usher et al. 1995; Frugal et al. 2003; Powell and Powell, chapter 6, this volume).

A fifth means to introduce stronger environmental justice dimensions to report-back protocols concerns addressing varying levels of environmental health literacy. Biomonitoring studies involve populations with varying levels of literacy (Boston Consensus Conference 2006). In some cases, as with the EWG breast milk study, participants are pooled from populations of environmental activists who already have high levels of environmental health literacy. One academic scientist we interviewed who was conducting breast milk biomonitoring noted that participants came from two distinct groups, one of which was upper middle class with a postgraduate level of education, and the other that was working class, with a high school or lower level of educational attainment. The latter group was far less inclined to seek their biomonitoring results. However, participants who are members of marginalized groups with low levels of scientific literacy may be eager to hear their results with a preference to have materials read to them and be shown diagrams, graphs, and pictures to interpret data (Quandt et al. 2004).

Another scientist involved in a cohort study on pesticides in low-income urban women and children provided further evidence that populations of low literacy are interested and can demonstrate a high level of comprehension in interpreting individual results: "Yeah, the research workers have been getting the same questions that they've been getting for years now, you know, when are we going to get our individual results for our kids? You know when are we going to know about pesticides? When are we going to know the results from our [monitoring]?"

Thus, successfully conveying complex results to populations with low levels of scientific literacy requires carefully crafting risk-communication protocols so that participants are engaged and able to understand the material presented to them. It is also necessary to communicate with members of the participant community during the creation of the report-back process, to ensure that the information is relevant to their life experiences. If these measures are successfully undertaken, there is no reason to assume that populations with lower levels of scientific literacy necessarily will have less interest in receiving their individual data than more educated groups. Ultimately, a participant's decision about receiving individual results is a personal one, and researchers must ensure that they can make a clear, deliberative choice regarding their right-to-know or not-to-know.

Sixth and finally, we would highlight the importance of negotiating individual-report-back objectives with IRB requirements and standards of confidentiality. Although IRBs focus on protecting the rights and confidentiality of individuals, this may not require that individual results be reported back to study subjects. In fact, under certain circumstances, IRBs may discourage individual report-back. The scientists we interviewed faced a range of responses from IRBs to their report-back protocols. One researcher recounted how the IRB initially opposed releasing individual study results to participants. However, he was able to convince IRB members to reconsider their decision by demonstrating that community representatives on the study's advisory board supported the report-back protocol. Another IRB limited researchers to calling participants and referring to them by their individual code number, rather than their names in order to protect confidentiality. Conversely, environmental advocacy organizations that conducted studies gave participants numerous opportunities to discuss their individual results with each other. In one study we examined, conference calls were held for all participants before and after results were disseminated and participants were encouraged to share their personal response to receiving their results with the group. This approach could encourage reevaluation of traditional protocols aimed at ensuring informed consent and protecting participant confidentiality and suggest new ways for researchers and communities to enhance the participatory nature of disseminating and interpreting biomonitoring results.

Payouts of Personal Exposure Science for Community and Scientific Partners

As discussed above, our interest in the scientific and ethical challenges of report-back from personal exposure research stems from our own research collaborative that involves environmental sampling of household air and dust in Cape Cod, Massachusetts, and Northern California to assess the presence of pollutants potentially linked to breast cancer and respiratory health outcomes. In addition to reporting aggregate exposure assessment results through peer-reviewed publications, media outreach, and public meetings, we also contacted study participants individually, and are carrying out a pilot intervention to reduce exposures.

Academic-community collaboratives are complex endeavors that require significant investment in building relationships to ensure that the goals, objectives, and needs of each partner are clearly addressed. In particular, the willingness of community-based groups to invest significant resources in the scientific enterprise depends on whether this work will advance their short- and long-term interests without straying from their primary organizational mission. Our community-academic research collaborative had its roots in an Environmental Justice Program funded by the National Institute of Environmental Health Sciences, with the funding supplemented by a grant from the National Science Foundation. One academic collaborator had a long-standing relationship with the Silent Spring Institute and another had a long-standing relationship with Communities for a Better Environment.

Our community partner, Communities for a Better Environment (CBE), has a demonstrated history of doing its own scientific work and leveraging the data it collects to push for policy and regulatory change. For example, the organization is well known for pioneering the "Bucket Brigades" for low-cost air sampling, used widely in California and Louisiana by fenceline communities living near large industrial facilities with hazardous emissions (Lerner 2005; Ottinger 2010). In the San Francisco Bay Area and Los Angeles, CBE has a long history of tracking and analyzing flaring activity and emissions from large oil refineries; this scientific work led to the promulgation of a groundbreaking flare control rule that became a front-page story in the *New York Times* (Marshall 2005).

For CBE, the Northern California Household Exposure Study was as much about outreach and community contact as it was about getting scientific results to push policy. CBE organizing staff were trained by Silent Spring Institute and university scientists to conduct the indoor and outdoor air monitoring, dust collection, and interview techniques, thereby enhancing their in-house scientific capacity and ensuring their co-ownership of the research process. Most importantly, CBE's partnership in the Household Exposure Study helped the organization demystify science for their constituents by enabling staff to move their data-gathering efforts into the realm of people's homes. For example, as CBE interviewers went through the preliminary exposure questionnaire and set up sampling equipment, the experience encouraged community members to think in new ways about indoor air quality and how contaminants from outdoor pollution sources can penetrate inside the home. These discussions enabled CBE to connect the socioeconomic and political aspects of their organizing work with the technical aspect, both of which are central to advancing environmental justice.

Decisions about what to report to participants were made collectively, following a model the Silent Spring Institute used in the Cape Cod study. In that model, findings were summarized in one page, and graphs prepared by public health and social scientists showing all detected compounds were enclosed. The graphs showed the levels detected for each participant's home, compared to levels found for all other study participants. Pollutant levels were also compared with regulatory benchmarks, where available. All collaborators divided up the task of writing the summary letters, which were double-checked for consistency and scientific accuracy. CBE staff personally met with participants to go over these materials and to help them think through individual and collective strategies for reducing those exposures.

Furthermore, this research enterprise, from the development of specific hypotheses to the report-back of study results, was designed and implemented collaboratively. This participatory approach draws from the notion of collaboratively doing, interpreting, and acting on science, a knowledge-production structure that is not linear, but rather cyclical, in that the collective process of acting on science leads to the further doing of science (Brown et al. 2006). We believe such a unified approach increases scientific quality, democratic participation in science, and the capacity for the science to effectively support education, organizing, hazard reduction, litigation, and policy advocacy.

The Household Exposure Study has some inherent methodological challenges, including a limited ability to speciate and link indoor and outdoor air pollutants directly to nearby refinery emissions. Despite this limitation, CBE has reported individual and aggregate sampling results. Its goal was to push the regulatory community and Richmond city planners to better account for the cumulative impact of multiple pollutants from industrial and transportation sources when making siting, permitting, and land-use decisions. For example, with the support of scientific partners, CBE, along with some study participants, has used data from the Household Exposure Study in its testimony before Richmond's Planning Commission to protest a conditional-use permit application by the nearby Chevron Refinery that would expand the facility's capacity to refine lower-grade crude oil and significantly increase its pollutant emissions. The presentation of the Household Exposure Study results received significant media attention, as well as inquiries from the California Attorney General's Office, both of which compelled the Richmond Planning Commission to allow for more public input on the environmental impact statement of the proposed refinery expansion. Ultimately, the City of Richmond approved the permit and the struggle went into litigation. A recent State Appeals Court decision on the case upheld a lower court's ruling that the environmental impact review for Chevron's conditional-use permit to expand its operations violated state environmental laws for being inadequate and vague about the scope and community health impact of the proposed project. The new scientific protocols were the result of transformations in practice instigated by EJ organizations and advocates for stronger participatory research structures.

CBE's engagement in the Household Exposure Study has brought important benefits to its scientific partners by improving the project's *rigor, relevance,* and *reach*. The scientific *rigor* of the Household Exposure Study was ensured through collective discussion and negotiation of study design issues, such as choosing relevant study sites, recruiting study participants, finalizing the list of chemicals for analysis, and developing sound protocols for reporting study results. For example, CBE encouraged the study team to collect a subset of air and dust samples from a community that did not have significant outdoor industrial and transportation source emissions so that these results could be compared to what was found in Richmond. Similarly, the *relevance* of the Household Exposure Study was bolstered through the collective development of bilingual (Spanish/English) graphic displays for communicating aggregate and individual-level sampling results to ensure that they were understandable and transparent, conveyed scientific uncertainties, and elucidated potential strategies for exposure reduction by highlighting emission sources. Finally, CBE's engagement in the Household Exposure Study helped extend scientific capacity to *reach* broad audiences in order to leverage results to improve regulation and land-use decision making. For example, scientists and CBE collaborated in the initial presentation of scientific results in order to train community partners to effectively use the data graphs and communication materials on their own at subsequent community meetings and for testimony in regulatory and policy forums.

Academic and Silent Spring Institute scientists also had the benefit of training postdoctoral fellows as well as graduate and undergraduate students in the social and public health sciences in community-based participatory research methods. The Silent Spring Institute, a leader in research on environmental links to breast cancer, used the Household Exposure Study to strengthen its linkages with communities of color working on environmental health and justice issues. Similarly, the collaborative enabled the organization to successfully inject an environmental justice perspective into the broader breast cancer community through its involvement in local fundraising, public education, participation on state and federal advisory committees, and involvement with national breast cancer organizations.

Because the Silent Spring Institute was already committed to community-based participatory research and environmental justice, the lessons learned for those concerns were not as dramatic as in other locations we and others have studied. A growing body of knowledge from our own interviews and from others (Fisher 2000; Shepard et al. 2000; Quandt et al. 2004) shows that the ruptures instigated by working on community-based participatory research projects with environmental justice groups have transformed the way scientists conduct research. These collaborations have given many scientists a broader awareness of how environmental health risks can be identified by laypeople, and a strong sense of the importance of community involvement throughout all phases of a study. These lessons are especially visible in talks and workshops at the annual grantees' meetings of NIEHS' EJ and CBPR programs, and they have been the focus of recent NIEHS-wide discussions and workshops. The growing application of CBPR methods in environmental health research (Morello-Frosch et al. 2006; Peterson et al. 2006) in the last two decades indicates an important shift in how such scientists view expert cultures. Moreover, these partnerships are helping establish best practices for reporting individual-level exposure data in the context of scientific uncertainty and political struggles over exposure reduction.

Conclusion

Biomonitoring and household exposure assessment exemplify how new techniques and innovation in environmental health science advance our collective capacity to detect and understand the health implications of chemical trespass in people's homes and bodies. Some of these personal exposure assessment projects are done by academic, government, and regulatory institutions, and involve varying degrees of lay involvement. Others are done by environmental health advocacy organizations in order to mobilize the public and lobby relevant officials and legislators for regulatory and policy change. These types of studies, many of which have their origins in EJ activism, also raise new ethical challenges that require democratizing the research enterprise so that participants can play a larger role in interpreting, disseminating, and leveraging study results to protect community environmental health.

Much of this scientific work involves informing individuals of their chemical exposures, and the proliferation of individual report-back approaches discussed here represents a departure from traditional models of reporting aggregate study results in ways that are limited to academic settings, such as professional meetings and peer-reviewed publications. Of note is the increased effort among scientists to report chemical exposures whose clinical significance may not be fully known. There is a need for guidance on the ethical responsibilities associated with communicating individual- and community-level data. Much of this guidance cannot come solely from the established arbiters of clinical and research practice, nor from government health officials, but must also include environmental justice and health advocate communities engaged in research collaboratives that have developed new standards of ethical report-back and democratic science practice. Our example from a multiparty collaboration shows how new CBPR approaches to biomonitoring and personal exposure assessment advance the goals of better science and a more empowered citizenry.

Acknowledgments

This research is supported by grants from the National Institute of Environmental Health Sciences (1 R25 ES013258-01), the National Science Foundation (SES-0450837 and SES-0822724), and the National Heart, Blood, and Lung Institute (T15 HL069792).

5

Middle-out Social Change: Expert-Led Development Interventions in Sri Lanka's Energy Sector

Dean Nieusma

Introduction: Middle-out Social Change

This chapter introduces the concept of *middle-out social change*, both as an alternative to top-down and bottom-up approaches to environmental justice and as an analytic lever to open up "expertise" and its potential role in democratic decision making. In top-down initiatives, experts typically work on behalf of policymakers in established institutions, enabling the development and implementation of systematic intervention strategies that are, to varying degrees, imposed on organizations, groups, and individuals occupying lower positions in the political hierarchy (Flyvbjerg 1998; Fischer 1990). In contrast, bottom-up or grassroots interventions are initiated outside of powerful institutions and typically entail demands for institutional change, including changes to the conceptualization or practice of expertise (Martin 1996; Sclove 1995). In grassroots movements, ordinary citizens are united by the shared perception of a common threat, allowing the consolidation of what was previously highly distributed political power (Laraña, Johnston, and Gusfield 1994), and, as part of that process, legitimating local knowledge is often seen as central to political empowerment (Wynne 1996; Fischer 2000).

Since grassroots groups can rarely afford to hire expert consultants at market rates, however, the role of experts in bottom-up change is not clearly demarcated (Woodhouse and Nieusma 2001). To fill the void, "expert activists" (Frickel 2004; Frickel, chapter 1, this volume) often work on behalf of grassroots movements or other marginalized groups for little or no pay, but in so doing they occupy a very different structural relationship to their "clients" than do traditionally employed experts (Hess 2007; Ottinger 2010). The more central the role played by expert activists in defining environmental interventions, the less sense it makes to describe them as "grassroots" initiatives, regardless of the extent to which the objective is grassroots empowerment. Middle-out social change accounts for the unique position occupied by expert activists, both in relation to the grassroots interests they purport to serve and the more powerful institutions typically confronted in environmental justice struggles. In so doing, it also exposes the exclusionary tactics commonly employed in expert interactions (Hilgartner 2000; Frankena 1992) and suggests strategies to overcome expertise-justified exclusion without denying the centrality of expert involvement in framing and solving problems facing ordinary communities.

Using a case-study approach, this chapter analyzes the activities of one organization—the Energy Forum of Sri Lanka—and the strategies it employed to achieve its particular version of environmental justice: equitable access to clean, affordable energy options. Uniquely situated between grassroots stakeholders and elite policymakers, the Energy Forum used its institutional placement to identify opportunities for bridging gaps in policymaking networks—both up and down the political hierarchy as well as horizontally with other midlevel experts and organizations. The Energy Forum's organizational expertise spanned multiple domains of knowledge, including renewable energy technologies, community organization and awareness raising, and national and international environmental policy. More important than its substantive expertise in these various domains, however, was the Energy Forum's ability to recognize and represent the perspectives of diversely situated stakeholders from across Sri Lanka's energy community, and in particular those of the rural poor (Nieusma and Riley 2010). The Energy Forum leveraged its substantive expertise to foster new forms of collaboration and representation, thereby transforming policymaking processes to insert community knowledge alongside more dominant technical and financial expertise.

The Energy Forum's overarching approach to development and environmental justice was not simply one of advocating grassroots participation, however. It extended to include advocating *a new model of interaction* among competing interest groups and the expertise at their command. Rather than focusing exclusively on bringing representatives of grassroots interests to the policymaking table, the Energy Forum sought to intervene in the very *mechanisms of exclusion* that made grassroots input moot in high-level policymaking even when grassroots interest groups did have a seat. In other words, through its various interventions—and situated as it was between the grassroots and powerful decision makers—the Energy Forum renegotiated the relative authority of various domains of knowledge in shaping Sri Lanka's energy and development policymaking spaces. It sought to transform the dominant decision-making processes by identifying and then modifying the institutional structures (and their underlying conceptual frames) that constrained effective participation by grassroots interests.

Highlighting the middle space between top-down and bottom-up approaches to environmental justice serves a variety of purposes. First, it adds nuance to what is otherwise a false duality in thinking about where social change originates: top-down and bottom-up change are not the only options available. In fact, nearly every social change advocate is located somehow between the top and the bottom of the invoked political hierarchy, and, as Schlosberg (1999, 90) points out, effective accommodation of diverse perspectives (i.e., pluralism) happens "in the spaces *between* individuals and the state." Second, the "middle-out" concept reminds us that experts occupy a variety of positions with respect to social change efforts—a central theme of this book. On the one hand, they represent a variety of interests: experts are not always in the service of the elite; many seek to represent marginalized groups. On the other hand, even those experts seeking to "represent" grassroots perspectives rarely do so without attempting to mediate, educate, or by some other fashion shift those perspectives: experts are change agents themselves. Third and most important, if addressed systematically, the middle-out concept directs attention to what I will term the *relations of expertise* that constrain typical grassroots empowerment initiatives by systematically excluding entire domains of knowledge. This analysis entails a critical orientation to expertise—one that recognizes the potentially productive role expert knowledge can play in serving grassroots interests without ignoring the social power embedded within expertise and, hence, its implications for democratic process. Ultimately, this chapter argues that, to be effective, experts practicing middle-out social change must accommodate the authority of expertise while simultaneously renegotiating how that authority is assigned and acted on by policymakers, other experts, and even grassroots stakeholders. In so doing, they not only operate in the (often unacknowledged) spaces between traditional, top-down expert practices and the grassroots activities commonly associated with the EJ movement; they simultaneously widen and make visible those spaces.

To build this argument, the following (second) section provides a brief overview of the empirical terrain—the context in which the Energy Forum worked—focusing especially on the organizations and interest groups that comprise Sri Lanka's energy and development arena. The third section describes in greater detail three categories of the Energy Forum's development interventions—down, up, and across the political hierarchy—showing how the middle-out approach was manifest in practice. The fourth section analyzes how these interventions serve to transform spaces of participation across Sri Lanka's energy and development community, particularly showing the Energy Forum's overarching strategy as one of disrupting dominant relations of expertise and seeking to reinvent those relations with more open, more democratic alternatives— alternatives that neither dismiss expert knowledge nor allow its (necessary) partiality to delimit the terms of deliberation in policymaking. Finally, the chapter concludes with some reflections on how the Energy Forum's approach might serve as a model for expert activism aimed at achieving greater environmental justice—socially, technically, and ecologically.

Setting: Sri Lanka's Energy Crises

An island nation off the southern tip of India, Sri Lanka (formerly Ceylon) has a population of around nineteen million people; enjoys higher than regional averages in health, education, and economic indicators; and has a relatively stable, democratic political sphere despite a decades-long internal war between government forces and the separatist Liberation Tigers of Tamil Elam (popularly known as the Tamil Tigers) that ended in May 2009. With the exceptions of the war and the devastation caused by the 2004 Indian Ocean tsunami, energy shortages have been the major development hurdle faced by the country since the 1980s. Since that time, electricity supply deficiencies have plagued Sri Lanka, resulting in substantial power shortages across the island for months on end. These shortages—"the energy crisis" as they are dubbed—capture national attention, and at their peaks surpass the impact of the war on people's daily life in the central and southern parts of the country, including the capital, Colombo. As electricity demand continues to grow and the national electricity provider, the Ceylon Electricity Board, relies increasingly on temporary thermal power installations to satisfy energy demand beyond existing generation capacity, both the country's generation deficiency and the environmental justice concerns associated with centralized fossil fuel energy facilities are expected to worsen. Besides the inequitable distribution of environmental burdens (e.g., pollution, noise) associated with centralized energy production, justice concerns extended to include questions surrounding the distribution of the benefits from investment in Sri Lanka's electricity infrastructure—the main topic of this analysis.

While the national electricity shortage captures front-page newspaper attention weekly for months on end, advocates for Sri Lanka's rural poor agitate for greater attention to "the other side of the energy crisis": over two million households are without access to grid electricity and its associated benefits. These households are forced to rely on low-quality, expensive, burdensome, and hazardous energy sources, such as kerosene for lighting and car batteries for TV or radio. Despite an impressive ascent in rural electrification rates over the prior two decades, by 2002 still only around 55 percent of the country's households had access to national grid electricity. Overlapping significantly with those promoting rural electrification is a community of renewable energy technology designers and advocates who promoted, designed, and implemented renewable, decentralized energy technologies, including solar, wind, small-scale hydro, biogas, and biomass (hereafter referred to collectively as "renewable energy technologies," or just "renewables"). While only a small fraction of the overall electricity generation mix, renewable energy is prominently figured in Sri Lanka. The renewables community is both coherent and diverse, including participants ranging from small-scale, independent technology developers to multinational corporations and projects ranging from those targeting poverty alleviation to those targeting climate change, and from locally designed, one-off hydro systems to standardized solar home packages sold by the tens of thousands. It was within this setting that the Energy Forum promoted its version of development and environmental justice: equalized access to clean, reliable energy for Sri Lanka's rural poor.[1]

A small nongovernmental organization (NGO) with a permanent staff of four to six members, the Energy Forum's organizational mission was to promote rural electrification as well as the adoption and dissemination of renewable energy technologies. Electrifying rural villages using renewables satisfies both goals, as well as the parallel overarching goal of "rural development." Formally registered in 1997, the Energy Forum is a spin-off organization of what was then called the Intermediate Technology Development Group (ITDG, now called Practical Action), an international development NGO founded by E. F. Schumacher and others in England in 1966. Over its history, the Energy Forum received funding from other national and international NGOs (including ITDG) and bi- and multilateral development assistance organizations, including the United Nations Development Programme (UNDP). As mentioned above, the Energy Forum's staff had formal expertise in renewables, community organization and awareness raising, and Sri Lanka's environmental policy. They also had informal, procedural expertise in bringing diverse actors together in productive ways—the topic of this chapter. Concerning environmental justice, the Energy Forum directed its attention to the injustices surrounding distribution and deployment of energy and development resources, and how those injustices manifest in particular types of communities, especially rural villages.[2]

As the Energy Forum worked to create social networks and policy contexts supportive of renewable energy and the needs of Sri Lanka's rural poor, it came into close and repeated contact with various other organizations and interest groups across Sri Lanka's energy and development arena—each with its own domain of expertise—including

• Government agencies, both national and provincial
• Bilateral and multilateral development organizations like the World Bank, UNDP, and the U.S. Agency for International Development (USAID)
• National and international private-sector companies, including distributors of renewable energy systems such as Shell Solar
• National and international NGOs
• Community-based organizations (CBOs)
• Energy end users or other intended beneficiaries of development programs, including the rural village dwellers commonly referred to collectively as "the rural poor"

Advocating renewable energy as an issue of environmental justice encouraged a range of changes in technical practice in Sri Lanka. Not only did the Energy Forum interact regularly and thoughtfully with a variety of players in each of these categories, but the organization went further to systematically reflect on those interactions, to devise alternative forums and modes of interaction that were more democratic than those already existing, and, ultimately, to implement strategies for altering the national policymaking terrain in order to formalize those more democratic modes of interaction and achieve greater development and environmental justice for all Sri Lankans.

Interventions: Working out from the Middle

Leaders of the Energy Forum conceptualized the various settings in which the organization worked as situated within a political hierarchy with "communities" (i.e., neighborhoods, towns, and villages as well as the CBOs that represented them) at the bottom; governmental and non-governmental organizations operating at the provincial, or regional, level in the middle; and national development policymakers—including national government agencies as well as companies, NGOs, and development assistance agencies with national reach—at the top.[3] The political hierarchy in which the Energy Forum operated reflects both the social networks in which the various players interacted and the relative power in policymaking of variously situated players across Sri Lanka. Precise placement of the various participants in a strictly hierarchical structure is less important to this analysis than communicating a sense of *varying degrees of influence*, with the Energy Forum operating as a national-level organization but without the influence of elite policymakers.[4] In fact, members of the Energy Forum understood their organization as situated somewhere between powerful top-level organizations and grassroots interest groups and individuals. It is the Energy Forum's middle positioning, and its correlating strategy to intervene across the entire sector from this positioning, that directs attention to middle-out social change and the role of experts in facilitating it.

To understand the implications of its middle positioning, this section reviews three types of interventions carried out by the Energy Forum in promoting renewables and the interests of Sri Lanka's rural poor—up, down, and across the political hierarchy. It will draw special attention to the way expertise is conceptualized and negotiated in order to set up the analysis in the following section. The first subsection describes the Energy Forum's approach to grassroots development using renewable energy technologies as well as what justified that approach. The second subsection considers key interactions with the World Bank in Sri Lanka and how the Energy Forum sought to open up World Bank development planning to a more diverse set of inputs. The third subsection describes the Energy Forum's horizontal work across the energy sector and its specific goal of creating a comprehensive network of organizations and interest groups spanning Sri Lanka's entire energy and development community.

Working "down" the Hierarchy: Interventions at the Grassroots

The Energy Forum conducted development projects in rural villages across the nation, with the exception of the conflict zones in the north and east. Grassroots development activities ranged from simple one- or two-day "awareness-raising" workshops to ongoing community organization activities to energy systems assessment and implementation.

Awareness-raising sessions were designed to educate rural villagers about Sri Lanka's national energy situation, environmental dimensions of the energy crisis, and renewable energy options for off-grid communities. Community organization activities, on the other hand, typically followed the implementation of particular types of new energy systems (including systems implemented by other development organizations) and sometimes entailed the formation of legally incorporated village-level "societies" that would oversee all aspects of energy systems management. The most ambitious type of grassroots intervention carried out by the Energy Forum was energy systems assessment and implementation, which entailed significant up-front research, awareness raising and appraisal of community needs and resources, and community organization, not to mention the technical design of new renewable energy systems. As an example of the ways in which the Energy Forum conceptualized its role and engaged grassroots community members, one project's "community needs assessment workshop" will be considered in greater detail.

As part of a larger project to pilot test community-scale dendro (wood-fueled) power plants in Sri Lanka, the Energy Forum identified a candidate site in the village of Bohitiya. An early step in assessing the feasibility of implementing the pilot plant in Bohitiya was a two-day community workshop, in which about 40 villagers (of a population of 1,400) participated. After introducing the organization and its mission and then reviewing renewable energy technologies, with an emphasis on dendro power, the workshop turned to highly participatory, contextualized community needs assessment activities in which the villagers energetically provided information to the design team, worked together collaboratively, learned to visually represent and document different types of information, and shared perspectives on the social and economic life of the village. The workshop closed with two activities: prioritization of Bohitiya's most pressing development problems and a review of the possibilities and limitations of the proposed dendro electrification project.

Across its activities, the Energy Forum sought to work side by side with the Bohitiyans from as early in the project development phase as possible and to share with the villagers control over project planning and implementation. But while the Energy Forum's project design team members were committed to the *ideal* of full community participation in project decision making, and even to community control over the project, they knew that implementing this ideal was far from straightforward. How could the team members hand over control of what they saw as an extremely important, extremely sensitive development pilot project—decentralized dendro electrification—when they also believed that the Bohitiyans' perspectives were (reasonably) biased by narrow, short-term interests (e.g., acquiring electrification) and a limited appreciation of the broader consequences of a failed pilot? Because they had no prior knowledge of dendro technology—either what it was or the resources required to operate it—the workshop participants were poorly positioned to assess the appropriateness of the technology and the larger development project it was part of. They also lacked the necessary information to situate the project in broader institutional, political, and technological contexts. Even with the information provided at the workshop, most Bohitiyans lacked the experience necessary to effectively assess the complex interactions among the project's various components, not to mention the diverse interests existing at multiple levels of the project beyond the local setting. For these reasons, and despite their ideals, the Energy Forum's design team members were unwilling to fully enfranchise the Bohitiyans with respect to the project—in effect justifying their own decision-making authority according to their determination of who held the appropriate knowledge.

Given this tension between their ideals and what they envisioned was responsible practice, the project designers struggled with the challenge of facilitating participatory processes in the village without jeopardizing the success of the dendro project. Bluntly put, the Energy Forum retained control over all project decision making including control over the *framing* of discussions about villagers' needs, aspirations, and experiences. The Energy Forum carefully crafted the format and objectives of the participatory activities and planned to continue doing so over the next several encounters with the villagers. This approach left villagers with the option of plugging in to the design team's structure or not participating. But the Energy Forum's power was not exercised in a heavy-handed way: team members did not impose the project or their perspectives on villagers. To the contrary, the very idea of community participation was motivated by openness toward, even recognition of the need for, the different perspective of the villagers and even a dose of dissent. Yet, the Energy Forum's team members recognized that the way they structured the workshop affected the ways villagers discussed and prioritized their "development" needs: whatever the dendro team members paid attention to, so too would the villagers.

Despite the real power differential, the Energy Forum's project designers explicitly sought to empower villagers, both cognitively and politically relative to other development actors (including the Energy Forum itself). Most centrally, the project designers engaged the process of community participation in the project as one of *mutual learning* through extended interaction. Since the designers knew that the way they structured their interactions with the villagers would impact the way villagers *understood* their needs, not just how they articulated those needs (Mosse 2001), they devised the intervention to give villagers: (1) knowledge about dendro electrification, (2) opportunities to evaluate the potential relevance of the project to their lives, and (3) the capacity to communicate their assessments to each other and to the Energy Forum. All the while, the project designers learned about Bohitiya, the quality of interaction among its various subgroups, and the conceptual, organizational, and material resources available in the village that might facilitate the project. Hence, learning occurred in both directions: designers learned how the villagers experienced life in the village and the conditions that gave rise to those experiences; villagers learned about energy technology options, new communication and representational strategies of their local knowledge, trade-offs associated with different energy-use options, and how others in the community understood life in the village. Villagers also learned about how their various development needs—such as electricity, economic opportunity, and community organization—were interrelated and would likely need to be addressed simultaneously. In this way, "community needs" were not simply *identified* by the project designers through community participation exercises. Rather, they were *constructed* through a complex mediation process that was orchestrated by the design team but that the villagers also played an active role in carrying out.

While the project designers accepted that differences of perspective rooted in different domains of relevant knowledge made simple, straightforward communication in either direction impossible, they operated under the assumption that through extended interaction, shared understandings could be created—understandings of both what the project would strive to achieve and how it would attempt to do so. Because such a shared understanding was a precondition for successfully transferring decision-making authority in the project, and because the skills learned through that process would benefit villagers in interactions with other development assistance organizations in the future, what was (simplistically) cast as a "community needs assessment" activity was, in fact, part of a much more nuanced strategy of empowerment. By explicitly recognizing and reflecting on their relative authority over the grassroots stakeholders (i.e., the Bohitiyans), the basis for that authority, and what it would take to transfer control over the dendro project to grassroots stakeholders, the dendro design-team members did not merely "assess" local needs but instead took local knowledge seriously in figuring those needs out. Despite maintaining control over project decision making, the designers did not simply override villagers' perspectives or attempt to supplant them. With their posture of mutual learning, the designers instead used community participation in the project as an opportunity to create bridges between various perspectives and their underlying knowledge domains. Through this process, the design team came to better understand the specific context of the proposed intervention and the villagers came to understand the requirements of that intervention. Both groups also learned to structure and communicate their knowledge in terms relevant to the proposed intervention that was located between them.

Working "up" the Hierarchy: Interventions within the World Bank

The Energy Forum's interventions up the hierarchy were of a very different sort than those down toward the rural grassroots, and the challenges of working up the hierarchy were nowhere more pronounced than with the World Bank. The World Bank's presence in Sri Lanka was notable for several reasons. Although it maintained a relatively small in-country staff, the nature and size of the development assistance programs offered to Sri Lanka meant international Bank staff members frequently visited Sri Lanka. Additionally, the Bank had contracted with a local Sri Lankan bank to administer its major renewable energy programs that were active in the 2000–2002 research time frame. The Energy Services Delivery (ESD) program was a five-year, $49 million credit line made available directly to private-sector actors for renewable energy production installations.[5] This program was scheduled to come to a close in mid-2002, but due to the program's success, according to the Bank's indicators at least, a follow-on program was implemented. The Renewable Energy for Rural Economic Development, or RERED, program picked up where ESD left off, this time providing an additional $134 million in development assistance for renewable energy projects.[6] While the Bank had other (nonrenewable) energy market development programs in operation in Sri Lanka during this time frame, the Bank's defining role in the renewable energy sector was of special concern to the Energy Forum. With over $180 million for investment over ten years, the World Bank was, financially at least, the sector's most influential player.

Along with financial resources, the World Bank brought to Sri Lanka's renewables community outside expertise, most notably renewable energy technology experts and development program design and management experts. This external expertise was in addition to the financial and program management expertise held by the Bank's in-country employees and contractors. Yet despite all this expertise, and despite the significant financial resources flowing through the Bank into Sri Lankan renewables projects, the Energy Forum (along with other organizations with extensive on-the-ground experience in rural communities) assessed the Bank's development programs as having serious shortcomings. Most notably, ESD was criticized for its utter failure to spur rural development activity beyond the direct exchange between technology seller and purchaser. This shortcoming was magnified by the fact that ESD's loan terms, especially for solar installations, meant rural community members ended up saddled with high-interest debt that many struggled to repay, thereby *reducing* their ongoing economic security rather than enhancing it. Different shortcomings existed with ESD's terms for microhydro systems. Since project developers would get paid in full on implementation and initial successful operation of their systems, there was no incentive to ensure that high-quality, long-lasting, low-maintenance systems were installed. Several microhydro systems broke down six to twelve months after implementation, with the developers already paid and departed, leaving rural communities forced to incur additional debt or go without the electricity created by the systems they were already committed to paying for.

The Energy Forum understood well these deficiencies of ESD and sought to rectify them by bringing its rural development expertise into the World Bank's development planning process—both during the final years of ESD and throughout the design of the new RERED program. The Energy Forum's general approach was to deepen and contextualize World Bank decision makers' perspectives on the renewable energy projects funded by ESD and RERED, providing an alternative reading of their "success" by shifting attention from high-level administrative and financial matters (e.g., number of systems sold, number of kilowatts installed) toward a range of impacts experienced at the grassroots. This approach was evidenced through a variety of the Energy Forum's interventions into ESD, including regularly representing grassroots interests in quarterly program stakeholder meetings, presenting systematic data on how ESD-funded projects were performing (poorly) on the ground in rural communities, and even bringing representatives of targeted communities to World Bank program review meetings when allowed to do so. These interventions were intended to call attention to facets of ESD that were otherwise ignored in the Bank's ordinary program management and assessment process, showing specific instances in which the program's purported success was misleading. However, a pattern of disinterest, and occasionally veiled antagonism, toward such interventions suggested the Energy Forum needed a new strategy for engaging the Bank.

As ESD came to a close, the design process for RERED provided new opportunities for interaction with the Bank. Given failures in shifting ESD, the Energy Forum changed its approach to RERED, moving away from providing alternative (read "competing") expertise in program assessment and toward what might be referred to as an adjunct model. Rather than positioning its role as that of a watchdog to the Bank's market-centered development approach, thereby setting up a power struggle the Energy Forum was certain to lose, the Energy Forum instead cast its rural development expertise as congruent with and supplementary, *but subordinate*, to the Bank's expertise in program finance and market creation. The Energy Forum's new approach was exemplified in its response to an early draft summary of the RERED program design that was circulated by the Bank. Instead of criticizing the new program for (again) failing to account for grassroots interests, the Energy Forum carefully responded to the document with a "constructive" (and upbeat) report suggesting a series of specific, highly focused modifications. These modifications were motivated by rural community experiences with ESD, but they were justified as needed to enhance the program according to the Bank's standard terms of assessment. In the most reductive terms, the Energy Forum argued that more confident and more satisfied customers would result in increased systems sales. The Energy Forum volunteered its assistance and expertise in implementing the changes and in managing the ongoing program. This model of collaboration met with better success, but only slightly.

One important concession made in the final design of RERED was a provision whereby microhydro developers were to be paid over a period of time following system implementation, rather than immediately on proof of operation, helping to incentivize installation of more robust systems and ensuring they were supported for some months into future. This change promised to improve outcomes both in terms of the World Bank's assessment process, which included some measures of project sustainability over time, and in terms of community impact. Other

Energy Forum suggestions for improving RERED, however, were not reflected in the final program design, and the Energy Forum was not invited to participate in program management beyond the existing public stakeholder meetings, either formally or informally. Worse, the Energy Forum never knew the specific terms according to which its suggestions were evaluated and, presumably, rejected, even though such rejection was not all that surprising to the Energy Forum given the history of resistance to its input by some of the Bank's staff, including especially the administrative managers of ESD and RERED. Resistance to the Energy Forum's input was particularly vexing given Bank representatives' repeated insistence that Sri Lankans take more "ownership" of World Bank–sponsored development programs.[7] In the face of such claims, the Bank retained strict control over RERED's design and implementation, including over the terms of deliberation surrounding program assessment.

The parallel between the Energy Forum's position relative to the World Bank and the Bohitiyans' position relative to the Energy Forum's dendro design team is notable. In both cases, concerns over "higher-order" development program matters overruled "local" perspectives and the situated knowledge that underlies those perspectives. As in Bohitiya, power differences (in this case between the World Bank and the Energy Forum) meant that the less powerful group had to justify itself according to the priorities (and expertise) of the more powerful group. However, there were critical differences in the two cases as well, differences aside from the fact that the Energy Forum was on the opposite side of the power imbalance. One important difference was the extent to which the decision-making process itself was reflected on and open to deliberation; another was the extent to which a posture of mutual learning undergirded that deliberation. World Bank decision makers not only appeared to be particularly unreflective about their exercise of authority or the institutional power and the narrow range of expertise that upheld it; they disallowed such deliberation and then, ironically, claimed "lack of ownership" at various local levels as a problem they faced. Worse still, especially from the perspective of democratic decision making, the Bank obscured the processes by which it evaluated external input, seemingly rejecting out of hand all of it that did not align neatly with its own development priorities and expertise.[8]

The Energy Forum, on the other hand, was highly reflective of its power—regardless of whether it held the position of lesser or greater authority—and stretched itself to learn to speak the languages of alternative perspectives and to respect the domains of knowledge that made them meaningful. The problem in interacting with the World Bank, then, was less in the Energy Forum's mode of collaboration and more in the Bank's overreliance on its own expertise in market development, its seeming inability to accommodate alternative domains of expertise, and its unwillingness to integrate divergent perspectives and to make visible the process by which determinations of "relevance" were made. The Energy Forum, in contrast, posited the relevance of any given domain of knowledge to a development program, and to the whole of "participation" in such programs generally, not as something self-evident but as a process to be deliberated and negotiated. And the Energy Forum came to this understanding in large part through the negative model provided by the World Bank. The benefits of engaging divergent positions, the Energy Forum learned, extended far beyond the completion—however successful—of any particular development project. The very act of deliberating priorities, negotiating competing perspectives, and responding to imbalanced power structures—and most importantly marshaling expertise in support of alternative courses of action—came to represent for the Energy Forum the core development challenge.

Working across the Hierarchy: Sectorwide Interventions
In addition to working down the hierarchy with less powerful community groups and up with more powerful international players, the Energy Forum also worked horizontally across the hierarchy, putting into conversation a variety of regional and national-level organizations and interest groups (and the diversity of expertise they employed) in Sri Lanka's energy and development arena. In addition, it attempted to integrate all levels of the hierarchy in a common conversation about Sri Lanka's energy future. This approach was motivated, in part, by a desire to achieve more effective deliberation across the sector, including better integration into existing networks of otherwise excluded perspectives and participants, especially those of the rural poor.[9] To achieve this ambitious objective, the Energy Forum created multiple contexts of moderated dialog among formal organizations and informal interest groups representing a variety of perspectives and expertise. It also created new organizations where they were most needed and where a critical mass of interest existed. Specific interventions at this level ranged from highly targeted, small-group activities to the widely attended National Energy Symposium.

One example of a highly targeted intervention was what the Energy Forum dubbed the "consensus meeting." The Energy Forum organized consensus meetings between various interest groups representing opposing sides of highly contentious energy-sector issues. One meeting, for instance, was organized between the Ceylon Electricity Board (CEB) and representatives of the "environmental lobby," which tackled entrenched differences over how to meet Sri Lanka's medium and long-term energy demand: pursuit of new coal-powered energy facilities versus decentralized, renewable energy technologies. The debate over coal power raged in Sri Lanka for years, being the frequent topic of national news headlines and regular newspaper opinion pieces. This debate was magnified in 2000 and 2001 because of the prolonged, repeated power outages caused by the energy supply deficiency. In the consensus meeting, the Energy Forum's coordinator carefully framed the boundaries of the dialog—"This is not going to be a debate. There will be no argument. We are here only to share each other's views"—and set up a moderated discussion, but otherwise played no role in advocating one position or the other.[10]

A similar meeting was organized between representatives of the World Bank and socialist provincial council ministers, which tackled the question of the proper role of government in promoting renewables and directing development assistance funds. In each of these meetings, the Energy Forum sought not so much to reach substantive consensus on areas of enduring disagreement, or even to resolve entrenched differences, but at least to create agreement that a more productive model of collaboration was needed and possible. In other words, the goal was to reach consensus on the desirability of a certain type of deliberative process if not its outcome.[11]

On the other end of the spectrum from interventions targeting specific interest groups or organizations, the Energy Forum sought to bring together many different interest groups—in fact, as many as possible—in a single, large-scale event: the National Energy Symposium. The first event of its kind in Sri Lanka, the National Energy Symposium was conceptualized, organized, facilitated, and sponsored by the Energy Forum, and it was the ultimate expression of the organization's strategy of fostering moderated dialog across the entire energy and development sector. The event comprised two full days of activity, including special lectures, thematic panel sessions, networking sessions, posters, a mini-tradeshow, and entertainment and awards. It was attended by over 500 participants representing the full range of energy-sector stakeholders, including national and regional government officers, representatives of the CEB and of renewable energy companies, environmental and community advocacy groups, finance and development agencies, energy researchers from all areas, and, most importantly according to the Energy Forum, energy end users from rural areas, including grid-connected energy users and those employing renewables such as solar, microhydro, and biogas. Also in attendance was the Minister of Power and Energy, the nation's top energy policymaker. To ensure diverse representation of interest groups at the NES, the Energy Forum secured attendance commitments from invited attendees, and, to facilitate the attendance of poor rural energy end users in particular, the Energy Forum offered an honorarium to defray travel and lodging expenses for those who required it.

While information sharing and informal networking were obvious goals of the symposium, the Energy Forum conceived of the event's main contribution as fostering partnerships:

The National Energy Symposium is designed to promote working partnerships that can effectively and fairly cope with the complex issues surrounding energy decision making. The goal is not to resolve all our differences of opinion but to openly discuss a range of viable energy options for Sri Lanka's energy future. Linking public, private, and civil society organizations is the necessary first step in creating such a working partnership. (Energy Forum 2002)

The importance of public–private–civil society partnerships was repeatedly articulated by the Energy Forum; however, forging such partnerships was not a trivial task given the power imbalances and the extent of the animosity existing among some members of the different sectors, as indicated above.[12] The National Energy Symposium was one way the Energy Forum sought to bring divergent interests and institutions to a common setting around a common theme—Sri Lanka's energy future—all with a celebratory ambience. But the NES also went further than advocating a strongly democratic, deliberative/collaborative model of energy policymaking; it instantiated that model by including energy end users as "equal" participants relative to energy policymakers (of all stripes), even if the equality achieved during participation in the NES was short-lived and primarily symbolic. Still, the symposium served the Energy Forum's agenda of reenvisioning Sri Lanka's energy sector as a whole, both by radically extending the number and types of direct participants in the conversation and by moderating the event in a way that suggested and implied all participants had an equal voice in—and relevant contributions to—the common conversation.

While the long-term impact of the NES is impossible to measure, it certainly achieved two of its objectives. First, with such broad-ranging attendance and the high level of enthusiasm among participants, the meeting effectively brought all major interest groups in Sri Lanka's energy sector into a shared and seemingly productive conversation—at least at that point in time. All participants, insofar as it was possible to discern publicly, seemed to view the event as a success. In his remarks at the event's closing ceremony, the Minister of Power and Energy congratulated the Energy Forum for taking the initiative to do what the government should have been doing but was not. Second, the Energy Forum integrated grassroots participants as equals—at least nominally—in national policymaking dialog that took place during the symposium. All participants enjoyed equal access to each of the sessions, creating a very different mix of perspectives than was typical in ordinary, by-invitation meetings. By its very structure, the symposium removed traditional barriers to participation, including those ostensibly based on expertise or "relevant knowledge," allowing for vibrant conversations among diversely situated participants intersecting in myriad ways. In a way, the symposium mirrored the Energy Forum's approach to reenvisioning the entire energy sector, prioritizing "decentralized, self-coordinated, and networked action," which is, according to Schlosberg (1999, 10), the hallmark of grassroots environmentalism. But the symposium itself did not arise spontaneously; it was achieved by deliberate (not exactly centralized but centrally coordinated) framing by a single organization—a tiny set of actors, in fact—with a relatively unusual understanding of expert interactions and the effective integration of expertise in democratic deliberation.

Transformations: Leveraging Expertise for Empowerment

Given the existence of wide-ranging opportunities for improving the integration of grassroots perspectives into Sri Lanka's energy and development policymaking, the Energy Forum's various interventions at each of the three levels described above could be viewed as a commonsense response to the managerial and political challenges facing the organization. I argue, however, that a more coherent, more sophisticated overarching two-pronged strategy was in play.

The first prong of the strategy was to bring the interests and knowledge of marginalized groups to the policymaking table, and a variety of the tactics described above worked to achieve this goal: representing excluded groups where they were absent; organizing grassroots groups so they could develop a unified voice vis-à-vis other stakeholders and so they could practice articulating and advocating their shared interests; and facilitating direct participation by grassroots community members whenever possible in energy and development policymaking. Though relatively straightforward conceptually, bringing marginalized groups to the table in an effective way required considerable effort by the Energy Forum. It required, first, establishing a shared understanding of each group's common interests and, second, committing to ongoing involvement to build the capacity needed to interact successfully with development program administrators and policymakers. Not surprisingly, the Energy Forum relied heavily on its expertise in rural development and its well-established relationships with numerous rural communities to execute this prong of the organization's overarching strategy.

The second prong of the Energy Forum's empowerment strategy was to ensure that, once they were at the table, marginalized groups' perspectives were actually accommodated. The Energy Forum's leadership recognized that participation by marginalized groups in development policy deliberation did not, by itself, translate into real influence, because their input was prone to facile dismissal as either uninformed or irrelevant (as was the case when the World Bank rejected grassroots perspectives in its program assessment meetings). Similarly, bringing together advocates of competing approaches to addressing Sri Lanka's energy crisis risked further entrenching well-established oppositional positions, as one group of experts rejected the grounds on which another group of experts stood (the precise outcome the consensus meetings were intended to counter). In both categories of dismissal, lack of appropriate expertise was used to a significant degree to justify the more powerful group's rejection of the less powerful group's input. And, importantly, this exclusion was not limited to grassroots (i.e., nonexpert) participants; development workers who could rightly claim expert knowledge in their particular areas of specialization were themselves marginalized by more authoritative experts, a practice justified according to the same fundamental logic. Not only were nonexperts marginalized by experts due to lack of "expertise" generally, so too were relatively less powerful experts marginalized by more powerful experts due to lack of "relevant" expertise (Nieusma 2007).

In the dominant "relations of expertise" existing in Sri Lanka's energy community, reference to (relevant) expertise, not raw political power, was used to circumscribe the influence of those participants representing less authoritative domains of knowledge, and this was done in the absence of deliberation over how relevance was determined in the first place. Even when participation by marginalized groups in decision-making settings was allowed, a kind of *epistemological exclusion* prevented their perspectives from being heard. Sometimes this exclusion was intentional—used by decision makers to overrule diverging input, as was likely the case with the World Bank's rejection of much of the Energy Forum's input. Sometimes it was unintentional—delimiting the terms of debate even as diverse input was ostensibly sought after, as in the case of the Energy Forum's reflections on its own power in framing Bohitiya's community needs assessment discourse. Either way, confronting epistemological exclusion was required if marginalized perspectives were to be heard, and this was the essential second step in the Energy Forum's empowerment strategy. Merely inserting diverse participants into the existing process was insufficient; confronting epistemological exclusion required reconceptualizing expertise as well as renegotiating its role in policy deliberations at every level of Sri Lanka's political hierarchy.

At each level of intervention, the Energy Forum's approach to empowerment recognized, at least implicitly, that expertise is something more than just specialized knowledge, existing somehow independently of the political and institutional contexts that give it meaning and, hence, authority. To the contrary, authority—that is, the power to overrule competing approaches or claims—was central to how the Energy Forum both understood and practiced expertise. In Sri Lanka's energy and development community, there were plenty of people with specialized knowledge that was arguably relevant to solving the development challenges faced in rural communities, but not all of that knowledge was understood to be expertise. Furthermore, not all of what was commonly understood to be expertise was equally authoritative relative to other domains of expertise. Instead, the authority of each domain of knowledge mapped onto its institutional placement in the broader political hierarchy, resulting in a "knowledge hierarchy" that closely overlapped with the political hierarchy.[13]

In this knowledge hierarchy, different domains of knowledge were afforded varying degrees of influence in policymaking, with expertise in finances and markets near the top (given, e.g., the World Bank's dominance in the sector), expertise in energy systems and agriculture somewhere in the middle, expertise in rural community organization lower still, and the "local" knowledge held by rural community members situated near the bottom.[14] Unlike political hierarchies, which are explicitly organized according to political power, the knowledge hierarchies operating within Sri Lanka's energy community tended to obscure the mechanisms by which one group exercised power over another. Using expert knowledge as the "admission requirement" placed the burden of countering exclusion on the victims of that exclusion rather than those who enforced it. By understanding that expertise entails authority—and that authority often translates into exclusion, whether deliberate or de facto—the Energy Forum's interventions went beyond mere advocacy of inclusion to address the mechanisms of exclusion operating within Sri Lanka's energy and development policymaking community.

Given the complexity of rural development, certainly many types of knowledge—financial, technical, agricultural, organizational, community—are relevant to rural development decision making. The Energy Forum's strategy for empowering marginalized groups sought to put various participants on a more equal footing in policy deliberations. Regardless of where in the political hierarchy it intervened, the Energy Forum strove to dismantle, with varying degrees of success, the knowledge hierarchy operating at that level: (1) it attempted to interject its own rural community development expertise into the World Bank's program-design process; (2) it created forums where a "discussion among equals" could take place in the consensus meetings; and (3) it adopted a posture of mutual learning in grassroots interventions that respected the knowledge of rural villages. In each of these cases, it should be noted, the Energy Forum did not seek to undermine the contribution to development work of any particular domain of knowledge; it did not seek to *invert* the knowledge hierarchy and, say, place rural development expertise above that of the World Bank's financial expertise. Instead, it sought to neutralize the role of the knowledge hierarchy in legitimating exclusion of certain stakeholders in policy deliberation.

Perhaps ironically, the Energy Forum leveraged its own expertise—in rural development especially—to counter epistemological exclusion justified on the grounds of other forms of expertise. On the one hand, the Energy Forum's long experience working with rural communities in Sri Lanka made clear that deference to expert authority could be assumed and that systematic effort was required to elicit genuine, thoughtful participation, especially in areas of potential disagreement or divergence of perspective. On the other hand, the Energy Forum's work with national development program policymakers, including the World Bank, made clear that deference to *their* expert authority was often expected by higher-status decision makers, and that systematic effort was required to open their decision-making processes to divergent input. In a way, then, the Energy Forum's positioning between the grassroots and elite decision makers allowed a perspective on the social power of expertise that proved invaluable in their understanding of the dominant relations of expertise at play in Sri Lanka, and hence in their challenges to knowledge hierarchies across the sector.[15]

The Energy Forum's alternative to the a priori assessment of the value of a particular perspective (determined according to its placement on the dominant knowledge hierarchy operating within a given context) was an open, deliberative process, where relevance of a particular perspective or domain of knowledge was determined through mutual assessment of its meaningfulness in context. Each of the organization's interventions can be seen as an attempt to "open up" the deliberative process existing at that level of interest-group interaction—not as a way of advancing a particular outcome, but as a way of ensuring variously situated stakeholders had a fair opportunity to convince others of the legitimacy of their input. Instead of employing appeals to expertise as a shortcut around democratic process, where experts decided on behalf of impacted populations what was in their best interests, the Energy Forum conceptualized expertise—in its various guises—as a contributor to democratic deliberation. In this way, the Energy Forum's process captured the productive potential of expert knowledge without allowing the (necessary) partiality of expertise to delimit participation or the terms of deliberation surrounding energy and development policy decisions.

By understanding how expertise operates as a combined force of knowledge and authority, and by reconfiguring how expertise was employed so as to counter inappropriate exercises of authority, the Energy Forum and its rural electrification work offered a transformative model of expert interactions. Its strategy required strong linkages among experts and between experts and nonexperts across a variety of communities of practice. It also required conceptualizing expertise not as an alternative to democratic deliberation but instead as a critical component of it. This model of interaction captures the potential of expert knowledge to contribute to effective policymaking, but does so in a way that does not preclude—and in fact demands—grassroots empowerment and participation in that process.

Conclusion: Middle-out Social Change as a Model for Expert Activism

Considering the Energy Forum's development interventions as a whole—down, up, and across the hierarchy—provides one set of answers to questions about how expert activists might productively direct their attention in promoting the interests of grassroots groups and others marginalized in policymaking. In the specific interventions delineated above, those answers revolve not around improved expert knowledge (as traditionally understood) in any one or more domains of expert practice, but instead around how expertise itself was assigned, understood, and performed.[16] Most centrally, that entailed renegotiating how various forms of expertise interacted within Sri Lanka's energy sector—that is, fostering "negotiation among the always partial and plural positions of professionals [i.e., experts] and lay people" (Corburn 2005, 41). Across the Energy Forum's interventions, and across Sri Lanka's energy-sector decision making, there existed a variety of participants representing a wide range of knowledge domains, cultural authority, interests, and perspectives. Rather than dividing the community neatly into the powerful elite served by professional experts and the marginalized rural poor with their local knowledge and, perhaps, a few counterexperts in the wings, the middle-out concept highlights a richer topography of knowledge, interests, and authority. It also highlights limitations in traditional conceptions of expertise as specialized knowledge abstracted from the social and institutional contexts that determine its relative authority vis-à-vis other domains of knowledge, expert and local alike. The Energy Forum's interventions can thus be seen as wedges into traditional practice, providing space to build new expert models.

If progress is to be made in addressing the complex challenges surrounding environmental hazards, development planning, or any other sociotechnical problem solving, the integration of various forms of knowledge and perspectives held by a diverse range of stakeholders is needed (Turner 2007). This chapter argues that what is at issue here is less the type of knowledge or the legitimacy of any particular type of knowledge and more the ways diverse types of knowledge are brought to bear on complex problems—that is, the relations of expertise. Better quality, more widely dispersed knowledge is surely desirable, probably even necessary, for better decision making, but it alone will not address deficiencies leading to sociotechnical systems injustices. Instead, attention should be directed to the processes through which diverse forms of knowledge (some, perhaps, incommensurable) compete for legitimacy in decision making and how such competition can be channeled into constructive dialog that does not cursorily exclude entire groups of stakeholders. Insofar as rural development—or any localized environmental justice initiative—is concerned, particular attention must be directed to how various forms of knowledge are made meaningful, in a wide variety of environmental policymaking situations, if they are to accurately and fairly represent on-the-ground conditions faced by burdened communities.

The concept of middle-out social change helps to reconceptualize environmental justice initiatives in the context of development by situating the locus of social change neither in parochial villages nor in elite-dominated development institutions and government bodies, but instead in the organizations, in the variously situated "experts," and in the relations of expertise that link the two. The Energy Forum's efforts in this vein strove to achieve a version of Schlosberg's (1999, 15) critical pluralism, which is interested "not just in equitable *representation* at the level of the state, but in ongoing diverse *recognition* and *participation* at a variety of levels and a number of practices of environmental policy-making, organizational politics, and community life."

Middle-out social change is poised to achieve this critical pluralism by situating experts *between* powerful decision makers (including but extending beyond those representing "the state") and marginalized grass-roots communities. In that in-between space, experts can play key roles, not merely in bridging the two groups, but in creating the new types of discourse and collaboration necessary to transport diverse perspectives and knowledge across the divide and throughout the network of interactions. This is not to say experts should be put in a position of final authority over development decision making; it is merely to take seriously expert agency in working between the grassroots and elite decision makers. In this way, expert activists can contribute to social change—and can even initiate it—leveraging their expert authority without falling prey to the myth of its ultimate authority in determining the best solution path to a complex social problem. The middle-out model of change requires us to critically evaluate the concept of expertise, but in so doing it allows us to reimagine the role of experts—and of expertise—in working toward environmental justice.

Notes

1. My ethnographic research in Sri Lanka was carried out between 2000 and 2002. Despite the ongoing work of the Energy Forum, I will limit my analysis to its activities through 2002 and, hence, use the past tense in describing the Energy Forum's activities.

2. Whereas environmental justice scholarship is typically focused on the unequal distribution of environmental burdens, such as polluting industries or other locally unwanted land uses, the Energy Forum was more concerned with protecting the environment generally through adoption of renewable energy

technologies and localized injustices in energy access deriving from resource distribution inequity. Certainly, EJ-styled opposition arose around the siting of permanent coal-fired power facilities in Sri Lanka (Nieusma 2004b), and the Energy Forum participated in such controversies in ways similar to those put forward in this analysis. However, as a researcher, I did not participate in interventions surrounding these controversies and so cannot detail the full-spectrum of "middle-out" activities involved. Nevertheless, a variety of "environmental" burdens also derive from the energy options available to off-grid communities in Sri Lanka. These are highly localized—fire hazards from kerosene lamps, waste problems associated with reliance on automotive batteries for household electricity, excessive crop damage from elephants in crops surrounding unlit villages— and not central to the Energy Forum's "justice"-motivated work. Switching attention from pollution hazards to less tangible burdens and from environment as background context to environment as active provider of needed resources is, I think, consistent with the spirit of EJ work taken broadly.

3. The Energy Forum's position regarding international organizations was that they should be considered an input to—not a determinant of—national policymaking, and hence were positioned apart from the "national level" in visual representations of the hierarchy delineated above. In practice, however, the top of the hierarchy was understood to consist of all elite policymakers—that is, all organizations with national reach, including governmental, corporate, NGO, and international development assistance organizations.

4. Schlosberg (1999) reviews the major criticisms of pluralism—and particularly its manifestation in the U.S. environmental movement's interest-group liberalism—that reflect inequitable influence. He offers "critical pluralism," derived from environmental justice activism, as an alternative.

5. ESD was unique in that it was among the first of the Bank's programs to provide development assistance funds directly to private-sector actors, instead of working through government agencies as per Bank precedent. Getting government "out of the middle" was a direction many Bank representatives in Sri Lanka wanted to move in.

6. RERED was unique in its own right. As a matter of policy, the World Bank did not invest in follow-on projects since the point of its assistance was to spur sustained development activity and follow-on projects threatened to foster dependence on ongoing assistance.

7. It should be noted here that a few employees of the World Bank, especially those coming from outside Sri Lanka, were sympathetic to the Energy Forum's input and perspective, but were not well positioned to advocate on behalf of that perspective.

8. In partial defense of the local World Bank staff, they were required to respond to institutional forces coming from the organization's international headquarters, which were outside my scope of analysis and outside the Energy Forum's scope of influence. This realization, however, does not address the lack of transparency of Bank protocols in the case of RERED or the lack of engagement by the local Bank staff with the Energy Forum, but it does limit the analysis of relevant

stakeholders to those operating within Sri Lanka, which is admittedly an incomplete set.

9. This effort resonates closely with Mary Follett's "attempt to define a unity that differed from uniformity" that is at the core of Schlosberg's (1999, 190) critical pluralism.

10. It was widely known that the Energy Forum was sympathetic to the anticoal position; however, the Energy Forum staff and directors also had effective working relationships with members of the CEB. In the consensus meeting, the Energy Forum's coordinator made clear that he would not be positioning the Energy Forum in the debate, and he asked me, as an "independent" outside researcher, to play the role of meeting moderator.

11. For additional detail on the Energy Forum's consensus meetings, see Nieusma 2007. Another type of example of a targeted activity was interest-group organizing, which entailed bringing together unorganized members of a common interest group and, where the potential for ongoing collaboration existed, capacity building to assist with the incorporation, vision setting, and ongoing management of the newly established organization. In particular, the Energy Forum sought to facilitate the creation of organizations where evident gaps existed in Sri Lanka's energy sector, gaps that needed filling for specific interests to be systematically represented within the sector. The Energy Forum spurred or assisted in the development of several such organizations, examples of which include community-based organizations (CBOs) in specific rural communities, organizational federations that connected geographically dispersed CBOs, and national tradesperson organizations for promoting the dissemination of specific renewable or decentralized energy technologies, such as biogas and high-efficiency cookstoves. Itself a recent spin-off organization of the Intermediate Technology Development Group–South Asia, the staff of the Energy Forum understood well what was needed to establish a new organization and get it off the ground, expertise that it made available to similarly situated advocacy groups.

12. The Energy Forum was committed to the ideal of the public–private–civil society partnership model, but it also promoted the partnership model for a strategic reason: to offer a counterpoint to the otherwise dominant neoliberal economic model. Given the privileged role of the private sector in the neoliberal model, which was promoted by international development assistance organizations, like the World Bank, the Asian Development Bank, and USAID and other bilateral agencies, the Energy Forum's advocacy of partnerships kept government and NGOs in the game.

13. The concept of knowledge hierarchies is elaborated in Nieusma 2007.

14. Although the knowledge hierarchy and the political hierarchy "closely overlapped," the overlap was by no means complete and neither was it universally held. In different areas of practice, the knowledge hierarchy manifest somewhat differently. For example, *within* the Ministry of Power and Energy, the World Bank, and the Ceylon Electricity Board, different domains of knowledge—local political culture, finance and markets, and technical respectively—enjoyed privileged status.

15. This point connects the Energy Forum's positioning and middle-out expertise to feminist standpoint epistemology (Harding 1991; Haraway 1988), where one's marginal position is not only another position among many, but provides an essential perspective on the operation of marginalization itself.

16. As indicated above, this is not to say that improved expert knowledge in a variety of domains of expertise is not desirable or necessary, and other Energy Forum activities pursued exactly that objective.

Part II
Extending Just Transformations of Expert Practice

6

Invisible People, Invisible Risks: How Scientific Assessments of Environmental Health Risks Overlook Minorities—and How Community Participation Can Make Them Visible

Maria Powell and Jim Powell, with Ly V. Xiong, Kazoua Moua, Jody Schmitz, Benito Juarez Olivas, and VamMeej Yang

Minorities and lower-income people are more likely to be exposed to a variety of environmental health hazards than white people—indeed, this is the essence of environmental injustice (Bullard 2000; Harris and Harper 1998; Lopez 2002; Mohai and Bryant 1992). Scientific and government institutions play important roles in constructing what we know and do not know about environmental risk disparities, and the ways these "knowns" and "unknowns" in turn shape scientific, political, and public attention to environmental injustices (Kuehn 1996; Stocking 1998; Wynne 2001). Institutional risk-assessment and communication approaches, typically embedded in Western European–based scientific cultures, are often blind to race, class, and cultural risk contexts and inequities, thereby rendering them "invisible" (Harding 1998; Harris and Harper 1998).

History shows that it takes community-based engagement from outside mainstream academic and government institutions to make race, class, and cultural disparities more visible in institutional risk assessments and communications (Cole and Foster 2001; Corburn 2002; Fischer 2000). In this chapter, we describe our work with the Madison Environmental Justice Organization (MEJO) to make these disparities more visible in public health agency risk assessments related to subsistence fish consumption. We describe how knowledge and communication gaps related to fish consumption risks are created and ignored by the same institutions that have power and responsibility for addressing them. We also highlight the obstacles MEJO has faced in bringing these gaps to light in institutional risk assessments, including: societal deference to what is perceived as more "valid" risk assessments of institutional experts, the parallel belief that local knowledge and community-based knowledge about the risks are less "factual" and therefore not valid evidence, and systemic indifference among institutional and political actors about class, race, and cultural contexts and how they are connected to environmental health risks.

While illustrating these obstacles, our engagement with government and academic institutions also reveals ways to overcome those barriers. By engaging diverse community members with institutional scientists in assessing and communicating fish consumption risks, we are bringing formerly overlooked knowledge and cultural perspectives of diverse people into institutional risk assessments, thereby creating productive ruptures in deeply ingrained institutional mindsets and scientific practices. The experiences of MEJO show that these ruptures, albeit incremental, are creating opportunities to slowly transform power relations among community members, scientists, and governmental institutions in ways that are bringing more attention to environmental health disparities in our community.

Our experiences suggest that institutional scientists can be personally transformed when they interact with people who have cultural backgrounds and experiences unlike their own. These interactions are encouraging some of these scientists to incorporate unique cultural and local knowledge of diverse community members into their technical risk assessments and communication strategies, in the process improving them substantially. Our projects, moreover, have demonstrated ways that power dynamics can be shifted in response to efforts of community activists working through institutional and political processes to push for change. In our case, MEJO has articulated race and class disparities in fish consumption risk assessments, gathered evidence on these disparities, and demanded through our empirical research and in public forums more power and voice in decision making on these issues. This is acting to prompt government scientists to work on improving environmental justice by supporting the evidence and proposals our group put forward and giving them their stamp of approval. Reporting from our perspective as participants in MEJO—in contrast to other cases in this book that analyze the intersection of scientific practice and environmental justice activity from the perspective of university-affiliated social scientists—our chapter documents how theoretically informed practitioners can help create ruptures in status quo scientific research, ruptures from which we hope to see new, more justice-based models in Madison grow.

Background: The Broader Regulatory and Legal Context of Fish Consumption and Environmental Justice

Subsistence fish consumption risks are unlike many environmental justice issues in that they involve risks related to activities people enjoy that can provide something healthy and necessary—food (as opposed to toxic waste dumps imposed on low-income neighborhoods). In the United States and throughout the world, fishing is important for people from nearly all cultures and racial and ethnic backgrounds (Beehler, McGuinness, and Vena 2003; Harris and Harper 1997; Whaley and Bresette 1994). Unfortunately, fish are contaminated with mercury, polychlorinatedbiphenyls (PCBs), and many other contaminants associated with reproductive, neurological, immune system, and developmental problems. Potential adverse effects are even more pronounced in children of mothers who eat fish during pregnancy (Buck et al. 2000; Jacobson and Jacobson 1996; Schantz et al. 2001).

Certain minority groups and poor people in the United States are more likely than others to rely on self-caught fish as a food source and therefore are more at risk from exposure to fish toxins (Burger and Gochfield 2006; Harris and Harper 1997; McGraw and Waller 2009; Schantz et al. 2001; Weintraub and Birnbaum 2008). Unfortunately, for a variety of reasons, these groups are also less likely to get the fish advisory information issued by state and local agencies and other organizations (Burger and Gochfield 2006; Powell 2004; Powell et al. 2007; Steenport et al. 2000).

Producing more data on race- and class-based fish consumption and communication disparities has been recognized as a critical environmental justice issue at the federal level for over a decade. President Clinton's 1994 executive order, "Federal Actions to Address Environmental Justice in Minority Populations and Low-Income Populations," includes specific requirements related to fish consumption:

Section 4-4 (Subsistence Consumption of Fish and Wildlife): In order to assist in identifying the need for ensuring protection of populations with differential patterns of subsistence consumption of fish and wildlife, Federal agencies, whenever practicable and appropriate, shall collect, maintain, and analyze information on the consumption patterns of populations who principally rely on fish and/or wildlife for subsistence. Federal agencies shall communicate to the public the risks of those consumption patterns. (Clinton 1994)

Recognizing the importance of comprehensive fish consumption information for all populations in order to do appropriate risk assessments and fish advisories, the U.S. Environmental Protection Agency (USEPA) issued the *Guidance for Assessing Chemical Contaminant Data for Use in Fish Advisories*. This document stresses that "selecting appropriate population exposure data is critical in both risk estimation and in fish advisory program planning" (USEPA 2000b, B-3), and that subsistence angling groups face potentially higher risks because of their higher consumption rates and "are at greater risk than the general population if their consumption is underestimated" (USEPA 2000b, B-3).

Executive orders and federal agency guidance documents notwithstanding, political priorities, policy decisions, and a plethora of technical and practical issues determine what kinds of monitoring and exposure data are actually gathered (Daughton 2001; Hess 2007; Kuehn 1996). Environmental media (water, air, soils, etc.), fish, and human exposure monitoring are typically responsibilities charged to state and local agencies under federal law. Monitoring is very expensive and time intensive. Unless these agencies have funding—along with political support, staff, and other resources necessary to do it—it will not happen. Unfortunately, the majority of fish and waterways in the United States are not monitored for synthetic toxins and, when they are, only one or two contaminants are usually assessed (Daughton 2001; Rosenbaum 2008).

Fish consumption data are even sparser. In addition, most consumption surveys to date do not include subsistence angling groups and/or racial/ethnic minority groups that are known to consume a lot of fish; most focus on sport anglers through fishing-license records (USEPA 2000b). This lack of data makes it difficult, if not impossible, to adequately assess risk related to fish consumption and to know which fishing groups should be priorities as far as communicating language and culturally appropriate fish advisory information.

As with monitoring fish and environmental media, assessing fish consumption levels often falls to state and local agencies that lack adequate resources and that are already overwhelmed by numerous other responsibilities (also see Hoffmann, chapter 2, this volume). Indeed, the EPA's fish risk-assessment document cited above suggests that "whenever possible, state agencies should conduct local surveys to obtain information on consumption patterns," but then qualifies the suggestion by noting that "the time and resources required to conduct onsite surveys . . . can be prohibitive" (USEPA 2000b, B-3). The document cautions, moreover, that surveys based on only those with licenses often underestimate consumption rates in some important fishing subgroups.

The EPA fish risk-assessment document, interestingly, suggests using qualitative, community-based strategies to get information about subgroups that are missing from standard consumption surveys. For example, it recommends that "anecdotal information is vital in directing the search for data on fish consumption patterns" (USEPA 2000b, B-5). For learning about specific locations where ethnic groups fish and/or for better estimates of consumption occurring via less direct or culturally specific routes (e.g., "informal" selling of fish, eating fish organs not typically consumed by European Americans, specialized cooking methods for ethnic recipes or rituals, etc.), it suggests that information be "acquired through informal discussions with local community groups in areas of potential exposures" (USEPA 2000b, B-6). Moreover, in describing potential strategies to learn about these issues and/or to reach important fishing subgroups, the document notes: "It may be most useful to enlist the help of local agencies or community groups to help access some of the subpopulations at high risk, such as urban-low-income populations or individuals of a particular ethnicity" (USEPA 2000b, B-8).

In sum, while fish consumption disparities are recognized as critical risk-assessment and environmental justice issues by federal agencies, in reality the local and state agencies that are in a place to gather more data to address and reduce these disparities are underfunded, understaffed, and often cannot reach racial and ethnic subsistence groups for a variety of other reasons. (Clinton's executive order, in other words, is a classic "unfunded mandate.") The EPA fish risk-assessment document suggests working with community groups to address these important gaps, because they are barriers to comprehensive risk assessments and fish advisory communications.

Our work engaging diverse community members with government agencies in addressing fish consumption risk disparities illustrates some of the concrete challenges in taking the steps mandated by Clinton's executive order and recommended by the EPA. Further, our projects reveal several deeper and more pernicious obstacles to bringing race and class risk disparities in fish consumption patterns to light—especially when these efforts are initiated by the community, *uninvited* by government agencies and institutions. At the same time, our case illustrates the opportunities and positive transformations that can be produced by creating ruptures in institutional cultures and scientific practices that are unlikely to be created in any other way.

Environmental Contamination and Subsistence Fishing in Madison, Wisconsin

Madison, the state capital of Wisconsin, has a population of about 220,000 people. It is the home of the University of Wisconsin–Madison, several other colleges, state government agencies, and cutting-edge biotech companies. It is a predominantly white, educated, middle- to upper-class community, and not typically viewed as a city with significant race and class problems. In recent years, however, growing numbers of Latino, Hmong, and African Americans have been moving to the city. These demographic changes have created increasing racial and socioeconomic disparities in Madison. A recent report, for example, states that although Madison residents "are more affluent . . . and better educated than their national counterparts, African American residents do not fully share this prosperity . . . and are significantly worse than the larger community in five leading indicators" (State of Black Madison Coalition 2008, v).

On the environmental front, Madison is renowned as a beautiful and "green" city. It is built around several freshwater lakes, called the Yahara Lakes. The Yahara Lakes have been central themes in the Madison area physically, culturally, and economically since the founding of the city in the mid-1800s, when the Ho Chunk (also called Winnebago) subsisted on abundant wild rice and fish from the lakes. Currently, the lakes are extremely popular for nearly all water sports, and are heavily and visibly fished by thousands of recreational and subsistence anglers from Madison and throughout the region.

In recent decades, however, several environmental problems similar to those of larger urban areas have been growing in Madison as the city's population has grown. The air quality is deteriorating to the point that Madison is close to being listed as a nonattainment area for particulate matter by the USEPA (USEPA 2006b; USEPA 2008) and asthma rates are among the highest in the nation (Warner 2004). Water quality in the lakes is worsening significantly as Madison grows, with fairly serious eutrophication problems from increased runoff and nutrient inputs.

Founded in this setting, the Madison Environmental Justice Organization (MEJO) is a small nonprofit, multicultural community organization that aims to combat environmental injustices in Madison by organizing the people facing environmental health risks to create capacity and leadership for change. One of MEJO's focal projects in recent years has been to work collectively with subsistence anglers, their families, and others who are concerned about fish toxins and water pollution to build awareness of fish advisories, and ultimately to reduce toxins in Madison lakes and fish so future generations can eat local fish without worrying about toxins. As part of its activities, MEJO involved people in research and discussion that offered alternatives to the conventional, government agency-related practices of risk assessment described above—creating a contrast between "street science" and "regulatory science" akin to that documented by Liévanos, London, and Sze (chapter 8, this volume).

MEJO formed when coauthor Maria Powell met Jody Schmitz, the food pantry coordinator at a subsidized housing complex in our neighborhood where many Hmong and other minority families live and regularly fish for food. Powell had recently completed a dissertation on fish consumption risk issues (Powell 2004), and fish contamination was a personal issue for her because she had grown up in Green Bay, Wisconsin, on the Fox River—a Superfund site (due to high levels of PCBs in water and fish). Schmitz was concerned that the people she served in her food pantry were eating a lot of local fish and did not know about fish advisories. Along with Schmitz and Powell, VamMeej Yang (Hmong Outreach Coordinator at the housing complex), Kazoua Moua (UW-Extension Hmong nutrition educator), and coauthor Jim Powell (community organizer) decided to organize to collectively address this issue. The group grew to include a diverse group of anglers and community members (Hmong, African-American, European-American, Chinese-American, and Latino), and we eventually decided to call ourselves MEJO. In 2007, MEJO received a small two-year U.S. EPA environmental justice grant to build community capacity to improve water quality issues through collective organizing, data gathering, and working with public agencies and university scientists. Ly Xiong was hired as the group's Hmong outreach coordinator.

MEJO's organizing is not part of any formal research project. Like participatory action research, our work aims to collaborate with community members in addressing community problems in what we decide together is a desirable direction (Gilmore, Krantz, and Ramirez 1986; O'Brien 2001). One key premise of our work is that all the community members we work with bring important knowledge to the organizing; another is that incorporating their perspectives and cultures will lead to more diverse and improved solutions to the issues we address. Each board member has experience and expertise critical to the work we do— including organizing, angling, fish consumption risks, cultural issues, languages, and more.

Minding the Gaps in Fish Toxins, Consumption, and Health Advisories

MEJO has worked to address problems in a community-based manner for Madison's growing minority populations, for whom fish are often important food sources: MEJO's surveys, meetings, and focus groups found that minority and poor anglers in the Madison area are likely to eat more fish than white anglers, making worsening environmental quality a particular threat to these groups. In addition, most of these subsistence anglers are not aware of fish advisories that would warn them about the risks of consuming fish from Madison's polluted waterways. In our work to assess and prevent disproportionate fish consumption risks to Madison subsistence anglers, we have encountered several critical "knowledge gaps" that need to be addressed (Frickel 2008):

- Gaps in data about toxins in Madison lakes' fish, especially the levels and kinds of toxins in particular species of fish
- Gaps in knowledge about fish consumption levels (i.e., what kind and how much fish people eat), body burdens, and potential health effects
- Gaps in communication and in knowledge about who fish advisories are reaching

Like other instances of "undone science" (Hess 2007; Frickel et al. 2010), these gaps are created by a variety of cultural and institutional factors (discussed further below) that ultimately stand as obstacles to environmental justice. Our projects bring together cultural knowledges and experiences to overcome those obstacles. The organization was founded on the belief that the best way to identify and fill risk data and communication gaps most relevant to the community is to involve community members in risk assessments and communications via collective organizing and collaborations with scientists, government, and other key actors (Brown 1992; Fischer 2000).

Madison Fish Toxin Data Gaps

Official agencies have little data on levels of toxins in Madison lake sediments, water, and fish. Figure 6.1 outlines gaps in fish toxin data. These lacunae are somewhat odd, given that Yahara Lakes are often touted as the "most studied in the world" because of the attention they have received from University of Wisconsin water science researchers. Government agencies in Madison and the University of Wisconsin are closely intertwined; many scientists who work in environmental and health-related government agencies have graduate degrees from UW and/ or are adjunct faculty there, and many university scientists serve on local, county, and state commissions related to environmental and public health issues. Government agencies in Madison often take their cues about environmental and public health issues from UW scientists and

	Mercury*	PCBs*	PBDEs	PAHs	Lead	Copper	Arsenic	Cadmium	Chlordane	Pesticide(s)	Pharmaceuticals	Other toxins	# of fish tested	# types of fish	Year(s) collected
Y = Data exists for this toxin / BLANK = No data exists for this toxin															
Lake Monona	Y	Y							Y	Y		Y	139	11	1985–2006
Lake Wingra	Y											Y	21**	5	1986–2004
Sugar River-Lake Belle View	Y	Y							Y	Y			13	4	1988–2004
Badfish Creek	Y	Y	Y#						Y				32	5	1990–2003
Token Creek-Rearing Pond	Y	Y											3	1	2003
Lake Mendota	Y	Y							Y	Y		Y	129	7	1983–2002
Lake Waubesa	Y	Y											69***	7	1976–2002
Yahara River-Downstream of Dunkirk	Y	Y											1	1	2001
Yahara River-Downstream of Lake Kegonsa	Y	Y											7	2	2001
Nevin Hatchery		Y										Y	10	1	2000
Crystal Lake	Y												10	1	1999
Lake Kegonsa	Y	Y											38	7	1976–1991
Yahara River-Below Stoughton	Y	Y			Y+	Y+	Y+	Y+				Y+	6	3	1991
Coliseum Ponds	Y	Y			Y	Y	Y	Y	Y	Y		Y	25	3	1990
Starkweather Creek	Y	Y											7	1	1990
Warner Park Lagoon	Y	Y							Y	Y		Y	8	3	1989
Graber Pond		Y							Y	Y			10	2	1989
Black Earth Creek		Y							Y	Y		Y	9	1	1988
Fish Lake Y	Y												14	3	1988
Grass Lake		Y											99++	1	1985
Nine Springs Creek		Y											17	1	1983

*For many samples, not all fish tested for this toxin

**For 6 fish, only mercury tested

478 — Fish tested over a 32-year period

#Only 2 fish tested for this toxin — = No data exists for this toxin

***For 41 fish, only mercury tested — = 10 or fewer fish ever tested for this body of water

+Only 2 fish tested for these toxins — = 5 or fewer kinds of fish ever tested for this body of water

++Only minnows tested — = Fish last tested before 2000

Figure 6.1

rely on their research to understand these issues—and this is especially true in regard to Madison lakes. Although recently a small group of graduate students studied toxins in sediments in a small area of one Madison lake (see below), we could not locate any University of Wisconsin research done in recent decades on levels of toxins in Madison fish (or people who eat fish).

A common reason offered by government agencies and university researchers for the lack of fish toxin data is that Madison lakes are not contaminated enough to be of concern for fish consumers, and that other areas (such as Superfund sites in Green Bay and Milwaukee) are far worse and should be priorities. However, Madison lakes (like all Wisconsin lakes) have been under fish advisory for mercury since 1983. The relatively few fish tested by the Wisconsin Department of Natural Resources (WIDNR) fall into ranges similar to fish in most United States freshwaters for mercury—0.12 to 0.47 ppm (USEPA 2000b). Madison panfish tend to be at the lower end of this range, while larger fish (bass, walleye, pike) tend to be at the higher end. For most fish in this range, consumers (especially women of childbearing age and children) should to some extent limit consumption to two to four fish per month. Some of the larger fish tested by the DNR had above 0.47 ppm mercury, and at this level, people are advised to eat no more than one fish meal per month.

Levels of PCBs in Madison lake sediments and fish are also well within the range in which people should carefully follow advisories. The Madison fish tested ranged from 0.05 to 0.46 ppm PCBs, and some fish had levels higher than this. The EPA recommends that to avoid "noncancer endpoints" (immune, reproductive, neurological problems), people should consume no more than half a fish per month with 0.19 to 0.39 ppm PCBs and no fish over 0.39 ppm PCBs. To avoid "cancer endpoints," the EPA recommends that people eat no fish that contain over 0.097 ppm PCBs, and only half a fish per month in the 0.048 to 0.097 ppm range—which would include all the fish tested in the Madison lakes (USEPA 2000b).

While these limited data suggest that levels of mercury and PCBs in some Madison fish are definitely high enough that fish consumers— especially sensitive populations—should restrict or avoid consumption, so few fish have been tested in the Madison lakes that it is difficult to adequately assess health risks to individuals who regularly eat this fish, or understand the scope of any health effects that might result from this consumption. Most contaminants other than PCBs and mercury have only been monitored in a few fish or are not monitored at all.

Raising some further red flags, a very popular fishing spot for subsistence anglers in the center of Madison, Monona Bay, was identified in 1987 by the DNR as a mercury and PCB "hotspot," and a later DNR report recommended more comprehensive testing of fish and sediments in that area to better inform fish advisories (WIDNR 2001). A small 2006 University of Wisconsin graduate student team project confirmed that Monona Bay is indeed a "hotspot" with several contaminants, including mercury, PCBs, polycyclic aromatic hydrocarbons (PAHs), arsenic, lead, copper, and zinc at high levels in sediment cores. Yet because there has been almost no testing of fish in Monona Bay, we do not know whether these sediment contaminants are accumulating in fish that anglers and their families consume.

Fish Consumption and Body-Burden Data Gaps

In addition to knowing what levels of contaminants are in fish, as the EPA's guidance document stresses, risk assessors need to know how much fish people actually eat and what levels of contaminants end up in their bodies. Only one very limited (and now outdated) study attempted to assess fish consumption levels among Madison subsistence fish consumers and subsequent hair mercury levels. In 1989, responding to questions raised by the Yahara Green Action Group, the Madison Department of Public Health (DPH—now a merged city-county department called Public Health Madison Dane County or PHMDC) surveyed 197 people (88 white, 68 Asian, and 41 African American) about their fish consumption levels and tested blood mercury levels of those who were willing.

The DPH study, published in 1991, was framed by the assumption that only very large fish in Madison lakes (like walleyes) would be of any concern—an assumption that was outdated a few years later when the safe tissue limit for mercury was revised from 0.5 to 0.05 ppm (National Research Council 2000). The DPH study was methodologically insufficient on a number of levels. Teenagers were hired and paid a commission to do surveys among "low income and refugees" (not defined), and it was later discovered that they "invented" some results. Only 27 of the survey participants volunteered to give blood for mercury testing, and they were not the participants who consumed the most fish. Based on blood from these 27 volunteers, the analyses found "no clear relationship between mercury blood levels and the number of sport-caught fish meals eaten per month," although three individuals had elevated levels of mercury in their blood. No race, class, or gender information about the hair sample volunteers was reported.

In conclusion, the DPH report recommended "follow-up testing and investigation on those individuals who exhibited elevated blood mercury" and continued evaluation of levels of mercury and other contaminants in area lakes and fish to "determine the need for future programming for high fish consuming groups." The recommendations were never followed, and in our experience of the past few years we would sometimes find this flawed study cited by public health officials as the basis for the premise that Madison subsistence anglers face minimal risks.

Communication Gaps: Fish Advisories and Media Coverage

Communication about risks with the people who are most vulnerable to them is a critical component of working toward environmental justice, yet this has not been pursued by Wisconsin state agencies with an eye toward those underrepresented populations. Such communication, though, plays a critical role in technical risk assessments in several ways. In the case of fish consumption, if subsistence anglers are not aware of risks related to eating fish, they are not likely to engage with community organizations or institutions working to understand and address them. Effective, inclusive engagement with anglers from diverse backgrounds, in turn, requires knowing about and respecting their knowledge about the physical, cultural contexts of fishing and fish consumption in their communities. Two-way communication with the most at-risk communities is necessary for accounting for the knowledge and perspectives of marginal populations in the risk-assessment process. It is also necessary for achieving the political solutions to address the risks so as to avoid further perpetuating gaps and disparities in risk assessments and communications.

Environmental and public health agencies in the United States and in the Midwest have issued fish advisories for many years (Tilden et al. 1997). These provide one way to reach fish consumers about risks related to eating fish and to offer advice about how much fish can be safely consumed. Unfortunately, studies indicate that fish consumption advisories tend not to reach the most vulnerable anglers and their families—particularly poor, minority, subsistence anglers (Burger and Gochfield 2006; Powell 2004; Powell et al. 2007; Steenport et al. 2000). These studies argue that one reason official advisories fail is because they are based on one-way communication approaches that do not take into account the diverse cultures, contexts, languages, and perspectives of the fish consumers they are trying to reach.

Language and translation are also important issues. Although both the Wisconsin Department of Health Services (DHS) and the DNR create fish advisories, only the DHS materials are translated into Hmong and Spanish. The DNR issues 1.6 million fishing licenses annually, but only prints 40,000 fish advisory booklets (all in English) (Schrank 2008). At the time MEJO's projects were initiated, there were no fish advisory signs along local waterways, including very popular shore angling spots in the middle of the city. None of the community members MEJO has worked with to date had seen agency advisories before MEJO brought them to community meetings; many expressed anger about not receiving fish advisories and said they would like translated advisory signs posted at shoreline fishing spots.

Mass media also play important roles building anglers' and the broader community's awareness of fish consumption risks and the political will to address them (Allen 2003; Burger 2000). In part reflecting local institutions' lack of attention to toxins in Madison lakes and fish, these issues have not been priorities for local media. Although declining water quality in Madison lakes has been a frequent theme of local media stories in recent decades, from 1989 to 2008 only 4 percent of 222 articles in the two local daily newspapers that were about the water quality of Madison lakes referred to toxins other than those associated with increasing nutrient loads in the lakes. Interestingly, one article in the early 1990s reported mercury levels in two local lakes as "among the highest in Wisconsin" (Associated Press 1990), while another a decade later reported that "fish [in these lakes] still don't get a clean bill of health" (Balousek 2000). The issue was not reported in the media after that. Mercury policy issues were periodically covered by local print media (186 articles from 1989 to 2008, per Lexis-Nexis database search), but only five articles during this time period connected mercury to locally caught fish and public health. All of those articles, in fact, were reporting on MEJO activities and were prompted by MEJO press releases (Weier 2006; Balousek 2008; Cullen 2008; Schneider 2008a, 2008b). Discussions about contaminants measured in Madison lakes and fish, moreover, have not entered into recent relatively high profile civic dialogs about "cleaning up the lakes"; the focus is on further reducing agricultural and construction runoff.

Participatory Research and the Madison Environmental Justice Organization (MEJO)

MEJO has been working to fill key fish consumption risk data gaps and to address environmental risk injustices by organizing with affected communities and doing community-based research. We organized community meetings and events to build community awareness of fish consumption, hear anglers' and other community members' perspectives on water quality issues, and engage community members in addressing them with us. In its first two years (2006–2008), MEJO held over thirty group meetings, organized ten "Let's Talk Fish" meetings, and held eleven public outreach events. MEJO has presented at several conferences, including the 2006 "Finding Solutions to the Global Mercury Crisis," an international conference in Madison concurrent with the Eighth International Mercury as Global Pollutant conference. MEJO members presented an outdoor workshop, "Minority Angling in Urban America," at Monona Bay during the conference, where sixty participants (including scientists from both conferences) ate a traditional Hmong fish dinner cooked by community members. Leaders in the group have organized more than ten meetings with MEJO members and public agency representatives and political officials. Our members and student volunteers have surveyed more than 275 people, primarily lower-income and minorities, about fish consumption and advisories in parks, along shorelines, in food pantries, at public meetings and events, and door to door. Of these activities, we elaborate on two—fish consumption surveys and the "Let's Talk Fish" events noted above—developed to fill risk data gaps and address environmental injustice,

The fish consumption surveys have been built as a participatory community-based data-gathering project. For several years, to fill a key data gap not being filled by institutions, MEJO members and volunteers have surveyed Hmong, Latino, African-American, and other anglers, gathering data about the kinds of fish they catch, buy, and eat, where they fish, how they prepare it, and how much they eat weekly. All of MEJO's surveys and most outreach materials are translated into Hmong and Spanish. MEJO members survey people in person at community meetings and events, at local food pantries, at local parks, and shoreline fishing locations.

MEJO members completed more than 125 fish consumption surveys, primarily among minority and/or low-income anglers (Powell and Powell 2008). Our data show that the levels of fish consumption among some minority, poor, and subsistence anglers range much higher than advisories recommend. Many minority survey respondents to date, for example, have reported eating fish every day or several times a week (recall that most Madison fish fall within two to four meals per month). Contrary to the assumptions of agency officials, a considerable portion of these anglers are not strictly eating panfish and other smaller, less contaminated fish. Many report regularly eating bass, carp, catfish, buffalo, walleye, and other larger fish that tend to have higher levels of mercury, PCBs, and other contaminants.

These results illustrated to us how important it is to assess *actual* fish consumption among specific groups and not assess risks based on averages or assumed fish consumption levels. Some groups eat far more than average. If risk assessors do not know the *range* of fish consumption among different anglers, they cannot identify those most at risk, communicate with them about ways to reduce or avoid these risks, or involve them in decisions about how to best address them (Burger and Gochfield 2006; Kuehn 1996).

Public health officials, to our surprise, are asking to see our data. Some have admitted that this is the only data they have on fish consumption in Madison. Further, the results of our data gathering motivated a DHS hair testing project at a community center. In a very small sample (ten people), hair mercury levels above the recommended limit were found in three minority subsistence anglers. MEJO is working with the health department and public health nurses to arrange to have further hair testing done in culturally acceptable ways. Akin to the report-back models pioneered by Rachel Morello-Frosch, Phil Brown, and their team of researchers (chapter 4, this volume) with respect to biomonitoring, MEJO plans to help agencies communicate hair mercury results sensitively and in the appropriate languages.

MEJO's collective work is also contributing to institutional fish risk assessments in ways beyond providing technical data, which, as the 2000 EPA document recognizes, is not enough to address risks and risk disparities comprehensively or appropriately. At ongoing community meetings, translated into Spanish and Hmong by MEJO members, we exchange stories about fishing and eating fish with anglers and their families. These informal conversations build trust and relationships with these communities, and also provide invaluable knowledge about cultural contexts of fishing and fish consumption. The exchanges are multidirectional; MEJO members from all backgrounds (including European-American) also share their fishing stories and cultures with community attendees.

In "Let's Talk Fish" meetings, for example, Hmong community members share stories about fishing in the high mountain streams of Laos and their pleasure in coming to Madison and finding many lakes and rivers with good fishing. Recent immigrants describe their adjustment to different styles of fish and fishing in Madison lakes versus high mountain

Laotian streams, and talk about Wisconsin fish that are similar to those caught in Laos. The meetings are also a space to share recipes, cooking styles, fishing stories, favorite fish, and whatever else people want to talk about related to fishing. Meetings include meals and are loud and lively social events, with lots of laughter and many children.

These interactions provide critical insights about cultural traditions and social contexts related to fish consumption that are unknown or overlooked in standard risk assessments about fish consumption among minority and poor communities. Most importantly, insights gained in these conversations further underscore how averages of population behaviors and lifestyles can erase particular needs of underrepresented and marginal populations, countering institutional risk-assessment assumptions that everyone behaves, eats, and communicates in the same manner (or like Midwestern Caucasians). For example, it is common for Hmong to eat the whole fish (sometimes including organs) in stews and soups and in certain cultural ceremonies. This is relevant for risk assessments because of biological considerations with regard to the distribution of contaminants in fish: some fish contaminants are more concentrated in certain parts of the fish than others. For instance, mercury concentrates in muscles and organs while PCBs concentrate in fat and skin. Assuming that people fillet fish and/or remove the skin, as is more common among European-American consumers (and recommended in advisories), likely underestimates the levels of contaminants potentially ingested in a fish meal.

MEJO also learned by talking with Hmong and African-American anglers that favorite types of fish among these groups are not listed on advisories. White bass, a favorite Hmong fish, and catfish, a popular African-American choice, are not listed. These omissions are problematic because white bass can have higher mercury levels, and catfish are likely to have more PCBs than other species. The brochure designers apparently assumed that these species are not frequently consumed and, having limited space, did not include them on advisories. This is a critical gap for people who eat these fish regularly, and would not have been brought to agency's attention without MEJO's work with Hmong and African-American anglers.

Through on-the-ground organizing and action with MEJO, we continue to learn a great deal about the cultural, socioeconomic, and physical-environmental aspects of fishing among different cultures, such as where, when, and why people from different backgrounds fish. Speaking with anglers along the shores of Madison lakes over the years, for example, MEJO members have learned that many African Americans drive several times a week from the inner city of Milwaukee to fish at publicly accessible spots in downtown Madison. These anglers, some of whom use fishing styles and traditions brought from the South, carry buckets of fish back to Milwaukee in their car trunks to share with friends and family there. Some of the fishers explained that they do not fish in lakes closer to Milwaukee because they are concerned about the racism in these communities. In 2005, for example, an African-American angler and his family were threatened with a gun and racial epithets by the fire chief and a firefighter in a community outside of Milwaukee while fishing there (Doege 2006; Kane 2005). In addition to the community-building exchanges that resulted from the "Let's Talk Fish" forums, this is one of many stories that frame the cultural views of underrepresented populations; these stories are important for risk assessment because they help risk assessors and communicators understand the sociocultural contexts in which people fish.

Government Institutional Cultures, Scientists' Choices, and Fish Risk Data Gaps

MEJO's participatory methods suggest the possibility of creating knowledge about fish consumption and fish advisories that can address the gaps that currently plague regulatory agency practices. Yet MEJO's research and organizing work have found that understanding and communicating risks to minority and poor subsistence anglers who fish from local lakes have not been priorities for government agencies and academics in Madison, despite the numerous political and scientific resources in this community.

The choices that actors at local and state levels—including individual scientists—make about what environmental health issues to study and act on, given political and funding constraints, are not deliberately ill-intentioned. Rather, they are rooted in long-standing political and socio-cultural values that shape institutional priorities; the government cultures provide the range of options available to the scientists. The individual actors, that is, work within deeper institutional structures of scientific research, with values so pervasive that they are as invisible as the minorities and poor they make invisible. Our work adds to EJ discussions about structural forms of injustice to suggest that institutional and cultural factors shape data and communication gaps in Madison through at least two avenues: (1) the fact that many scientists and key actors within government and academic institutions are not aware of and/or do not understand race and class disparities in environmental health risks, so they are invisible in their research and risk-assessment and communication strategies (a form of invisibility that resonates with the case Liévanos, London, and Sze -present in chapter 8 of this book when discussing California's Department of Pesticide Regulation); and (2) the dominance of and deference to institutional scientific risk assessments, and parallel discounting of community members' localized, contextual knowledge and experiences related to the risks. Combined, these factors exacerbate data gaps by creating chicken-egg problems: no toxin data, hence no risk, and no risk, so no need to get more data. They also constitute important obstacles to MEJO's efforts to advance participatory, culturally sensitive approaches to risk assessment, as evidenced by the examples discussed below.

Invisibility of Race and Class Contexts
How can very visible minority anglers be so "invisible" to academic and government scientists and officials? Many minorities fish daily from highly visible shoreline spots in Madison, just blocks from government agencies responsible for fish risk assessments and advisories. When asked why there are not more data on fish contaminants or consumption levels among Madison anglers, a common answer from government scientists we found was that "nobody eats very much of the fish" and/or "they only eat the small fish." Yet there is little evidence for these assumptions.

MEJO's interactions with agency and academic actors suggest that the apparent invisibility of these anglers is rooted in deep institutional "blindness" to Madison's race and class disparities, as well as to the diverse cultures and contexts of nonwhite people in the community. Government institutional cultures, we found, do not "see" minorities and poor for reasons that mimic long-standing patterns of institutional racism in the United States.

MEJO's efforts to convince local and state agencies to post shoreline fish advisory signs illustrate the blindness to race and class disparities among institutional risk-assessment professionals. In 2006, for example, MEJO attended a county Lakes & Watershed Commission meeting to ask for advisory signs along Monona Bay. Although we asked for Hmong translation well before the meeting and it is required by law, it was not provided. At the meeting, a Hmong MEJO cofounder testified on behalf of his community, saying that they would like fish advisory signs in

Hmong along Monona Bay, where many Hmong fish. Other MEJO members asked for signs along popular shore fishing spots in English, Hmong, and Spanish. The commission chairperson proposed a working group to develop a protocol for posting signs (Novak 2006).

Months passed, and after repeated MEJO queries, county officials attempted to convene the working group, which was to include several agency public health officials and a MEJO representative. The MEJO representative suggested that some minorities from outside of government agencies be included in the working group, perhaps anglers and/or leaders of minority health groups. She argued that including people of color on the task force would be the best way to bring their perspectives into the discussions about what should be on the signs and where they should be, and would consequently greatly improve fish communications with minority anglers in the future. The commission chairperson, however, was firm, saying that no one else could join the working group because the membership "was set." In the end, the group never met. The chairperson's refusal to diversify the working group suggests that county officials did not understand the contributions of people of color to be valuable or necessary to deliberations over fish advisories.

One year later, a county supervisor, the only person of color on the county board, contacted MEJO after seeing an article about our work in a local minority paper by one of our volunteers. That supervisor expressed interest in submitting a county resolution to install fish advisory signs. With MEJO members, the supervisor cowrote a resolution that described disparities in fish consumption and advisory awareness and required the county to place advisory signs in Spanish, English, and Hmong at popular shore fishing spots. The resolution also required that in developing the signs, agency officials work with environmental justice organizations and communities of color to determine where to post the signs and what should be on them.

The county officials' reception of the resolution further illustrates the problem of invisibility in risk-assessment and communication practices. In this case, shortly after the supervisor introduced the resolution, county agency staff drafted a substitute resolution that removed almost all of the original language (374 of the original 426 words). Their resolution removed the terms *minority*, *low-income*, and *of color*, plus all the text about data gaps and unknowns, and replaced them with text from general state fish advisories. Moreover, the substitute resolution significantly weakened the action items in the original resolution (see appendix I).

Weeks later, at another Watershed Commission meeting, both the original and the substitute resolutions were debated. Several MEJO members testified, including a Mexican-American angler who spoke about Latino cultural perspectives on fishing and lack of awareness about fish advisories in his community. Other MEJO members highlighted studies around the country that demonstrate race and class disparities in fish consumption and advisory awareness.

After hearing these impassioned testimonies, commissioners discussed whether to adopt the substitute amendment or keep the original. During this discussion, a member of the commission who is an emeritus professor, an internationally renowned limnologist, and a local watershed authority said he favored the substitute resolution without the social and environmental justice language because the watershed commission "doesn't deal with social justice issues . . . we just deal with water quality issues." The substitute resolution was quickly adopted by the commission without further opportunity for MEJO to comment. Similarly, at a later meeting with public health officials—one at which minority anglers again spoke about disparities in fish advisory outreach and awareness as important ethical and racial justice issues—a high-level state public health official and renowned national expert on fish consumption advisories said he did not want to bring environmental justice into this because it was "just a communication issue."

Finally, in April 2008, the county board passed the significantly weakened substitute resolution with all race- and class-disparity language and mentions of data gaps removed. The resolution described the installation of signs as a possibility, requesting a report on the issue from staff in three months.

In sum, MEJO's interactions with agency experts—many of whom are the key actors in government decisions about fish consumption risk-assessment and communication strategies—revealed the agency experts' limited understandings of the connections between racial and class disparities and environmental health issues. It is notable, for example, that two established white professional agency representatives (senior-level scientists with doctorates) felt comfortable explicitly stating in public meetings that race and class disparities in fish advisory awareness are not relevant for the watershed commission or the public health department, especially with people of color present who testified about their community's concerns about water pollution and fish toxins, and their lack of awareness of fish advisories. That racial disparities remained invisible to agency experts presented a significant obstacle to

MEJO's attempts to see the knowledge of minority groups represented in agency decisions.

Dominance of and Deference to Institutionally Sanctioned Scientific Risk Assessments

By working in the space where marginal communities encounter state agency scientific protocols, MEJO's research and activity highlighted a second structural obstacle to changes in risk-assessment and communication strategies: high trust for and deference to expert assessments among agency decision makers. Further, experts' status in this arena perpetuated the problem of knowledge gaps, even in light of participatory efforts to fill those gaps. Where institutional scientific experts had not measured fish toxins and consumption and/or had not brought these issues to public attention, decision makers assumed that they must not be problems.

This reliance on scientists to identify risk issues is a common theme in community-based struggles to address environmental risks:

Because the existence of such risks—let alone their origins and consequences—must be deduced by active causal interpretation, they exist in the social world only insofar as there is scientific awareness of them. At every stage in our understanding of such risks, the mobilization of scientific knowledge is central to their description and assessment. This elevates the expertise and status of the knowledge professions to a prime political position in the discourse of risk, leaving little or no room for the layperson. (Fischer 2000, 51)

MEJO's experience makes it clear that Fischer's insight applies not only to the way scientists and their official scientific data define fish consumption risks, but also to the role that the *lack* of scientific data plays in risk assessors' and community organizations' capacities to understand and address fish consumption risks. Lack of data contributes to the perception that there are minimal risks from eating Madison lakes fish, and then ironically, this perception is a key reason for the low priority placed on getting more data. High deference to institutional scientists' data on fish risks—perceived as more "valid" than other data—along with reluctance to recognize data gaps in institutional science, play pivotal roles in this chicken-egg feedback cycle (Hoffmann-Riem and Wynne 2002; Smithson 1989).

Specifics in the Dane County fish advisory debates illustrate how scientists' high status can preserve knowledge gaps and, in turn, justify inaction on environmental health concerns. Omitting statements in the original county resolution related to unknowns about environmental impacts of pollution on low-income and minority citizens, lack of data collection, and lack of consideration of these populations in determining public policy, for example, reflected authorities' reluctance to acknowledge that they had not researched these issues. Furthermore, changes made in the specific requirements of the resolution suggested not only reluctance to take action on posting signs but also unwillingness to credit conclusions reached by studies other than their own. Instead of requiring, as the original resolution did, that relevant agencies work with environmental justice groups and people of color to post and maintain fish advisory notices, the substitute resolution requires that agencies "*investigate* existing outreach efforts advising anglers in English, Spanish and Hmong" (emphasis added). The agencies were not convinced that there were any race- and class-based disparities in outreach efforts to date, even though MEJO proposed the original resolution *because* their extensive participatory research and outreach efforts had already established that minorities and the poor in Madison are less likely to receive and/or to be aware of fish advisories.

Again, MEJO's evidence in this regard, paralleled by numerous studies around the United States, did not match the institutionally accepted model of proper "scientific" practice; agency scientists felt they needed to verify it themselves with an "investigation" before they could justify the need for signs. This ended up delaying actions to post signs, the clear intent of the original resolution. Ironically, the agency health official charged with investigating existing efforts to communicate with minority anglers found and reported that MEJO had done most of the active, in-person fish advisory outreach to minority angling groups to date, and the only outreach in Hmong and Spanish (Public Health Madison & Dane County 2008).

These interactions illustrate institutional assumptions about who is qualified to define what counts as "valid" risk assessments and who is not, and agency experts' struggle to be in control of how fish risk issues are defined and communicated. Taken together, we saw these devalue the potential contributions of MEJO's participatory research. Furthermore, our interviews with a county official revealed that staff deleted the words *minorities, low income*, and *of color* in the resolution and replaced them with general fish advisory text because they felt that "fish advisory information is 'factual,'" while text about race and class disparities is "opinion." In so doing, the staff concealed and made invisible the values underlying their activities. Yet fish advisories, of course, are hardly accepted scientific "facts"; they are just as contingent on values and politics as other scientific information. Advisories have been debated by scientists for years, vary from state to state and among different government agencies, and have changed considerably over the last couple of decades as more regulatory scientific studies on fish toxins have been completed (Kuehn 1996; Moore 2003).

This is not to say that the race/class disparity text in the original resolution is any more factual than the advisory information that replaced it. It is, rather, to point out the irony that although both the original disparity text and the substitute fish advisory text are based in part on scientific studies, agency staff felt the substitute text was more "factual" than text submitted by MEJO and the county supervisor. Moreover, the experiences, knowledge, and cultural perspectives of the minority anglers who testified in these public meetings from the position of their communities' fishing cultures seemed to carry little weight in the discussions. The fact that these testimonies were not relevant to public officials was disheartening and demoralizing to MEJO members, especially the minorities who testified, and further demonstrates that institutionally sanctioned expert knowledge is considered more valid in fish risk assessments than community members' knowledge and experiences.

In sum, MEJO's membership challenged, and continues to challenge, conventional scientific means for crafting fish advisories while shedding light on the durability of institutional structures to resist those challenges. Indifference to race and class disparities by the public officials that represent agency expertise is less a reflection of purposeful individual indifference and more an indication of the deep deference to perceptions of the more "valid" scientific expertise of institutional scientists. It also brings to the fore the parallel belief that localized knowledge and experiences are less "scientific" and therefore less valid as sources of evidence (Fischer 2000; Gieryn 1999; Harding 1998; Irwin and Wynne 1996). Public officials' discounting of the anglers' testimonies and MEJO's participatory community research, moreover, suggests that officials are not yet convinced that Madison has race and class fish risk disparities, in part because academic and agency scientists have not validated that. In discounting minority anglers' cultural knowledge, agencies overlook essential information for appropriate and relevant risk assessments in minority and poor communities.

There are difficulties in shifting scientific models that our experiences brought to the fore. Seeking to reshape expert models in our grassroots projects has illustrated the frictions and cultural tensions that come with organizing against deeply entrenched institutional models. In one sense, explicitly recognizing such risk disparities would decentralize institutional scientists' roles in defining and managing fish consumption risks and undermine the premise of universal, abstract, and objective science. Perhaps more importantly, recognition of race- and class-based disparities, and the cultural factors that shape these disparities, would suggest that meaningful participation in risk assessments by minority and poor groups that eat the fish is essential, because they carry with them personal knowledge about their cultures, fish consumption, and advisory awareness in ways that institutionally based agency expert scientists cannot.

Conclusion: MEJO's Community Collective Work Transforming Institutional Risk Assessments

MEJO's ongoing work builds on environmental justice efforts elsewhere (including the many projects described in this book) in which laypeople and community organizations push from the bottom up to bring environmental health risk gaps and disparities to light in institutional risk assessments and policies. Our projects are creating productive ruptures in long-standing institutional scientific practices by bringing diverse people directly into risk-assessment and communication processes that typically privilege institutional scientific experts and that rely on abstract, reductionist approaches that tend to overlook race and class disparities as well as local cultures and knowledge.

MEJO has been slowly bringing the cultural contexts and knowledge of local angling communities to the awareness of institution officials and academic scientists by sharing these contextual factors in part through the various community interactions discussed above. The organization has also held several public events to raise awareness about fish contamination and the need for signs. It has also brought environmental justice in Madison to the public eye by sending out numerous press releases about these efforts, resulting in coverage in the local daily, weekly, and minority newspapers, as well as local TV and radio (Cullen 2008; Novak 2006; Schneider 2008a, 2008b; Weier 2006). The press releases and reports have generated media attention to the fish consumption disparities, increasing broader community and political awareness of these inequities. Such public events and media coverage play important roles in building public and political awareness and dialog about environmental justice issues that are otherwise invisible.

Coordinating these public outreach elements is part of, not distinct from, the work to reformulate scientific processes of risk assessment. All along, MEJO has facilitated the inclusion of minority anglers, leaders, and other community members in public meetings to share their perspectives directly with agency representatives, politicians, and scientists. Based on public officials' reactions in meetings, it is clear that many of them have not interacted with poor minorities in their professional work before and some are being exposed to race, class, and cultural perspectives they have never considered before. This very exposure at a cultural community level, we argue, is a necessary step toward changing risk-assessment procedures to include the experiences of affected populations.

Although there is no shortage of evidence of resistance to change and deference to entrenched models of risk assessment that will be difficult to transform, MEJO's work also shows signs that its efforts are slowly paying off. There are indeed small signs that these activities are changing the public officials' approaches to addressing these disparities. These include an increasing willingness to accept MEJO's data, collaborate with MEJO in gathering more data, and accept cultural and contextual knowledge of diverse angling communities as valid and important components of risk assessments. MEJO's efforts regarding fish advisory signs along local lakes have also encouraged public and policy discussions about risk and communication disparities. The advisory resolution that ultimately passed, although weakened, requires public agencies to work with MEJO to investigate communication efforts, which will hopefully engender further multicultural dialog about environmental justice in Madison.

MEJO leaders and members are developing working relationships with public officials and knowledge of political processes related to public health assessments, helping them become active participants in these processes. Several MEJO members have gained organizing experience and enough knowledge about fish consumption risks to be effective public communicators of their community's and MEJO's concerns. This in turn has helped MEJO's credibility with other local community organizations of color, and the group is beginning to develop collaborations with these organizations.

Multicultural organizing, of course, is extremely time intensive, as well as politically and culturally challenging. In part because of existing segregation, racism, and lack of access among minority groups in Madison, groups of color have limited power in the community. Anger and tensions about this lack of power and access, and about deep and systemic inequities between minorities and whites, at times create emotionally charged public and political interactions. Moreover, although MEJO has begun to bring Hmong, Latino, and African-American groups together—an important accomplishment in itself—it has been difficult to facilitate sustained participation by people from different ethnic and racial backgrounds. Cultural and language differences among racial/ethnic groups can be pronounced, and are sometimes barriers to effective collective organizing.

Beyond the essential step for creating increasing awareness of environmental injustices, the case of MEJO illustrates mechanisms that can gradually transform institutional scientific practices—such as creating common meeting forums, working to negotiate language with various stakeholders, diversifying voices in relevant debates—that we hope will provide examples for scientists, scholars, and activists alike. Our work illustrates some ways that risk scientists and communicators can effectively engage with diverse people affected by risks, and incorporate their knowledge into risk assessments—making these assessments not only more equitable and culturally relevant but also more comprehensive and accurate. Risk communications based on these improved assessments, likewise, will be more just and relevant and will hopefully reach people not previously reached because risk assessments rendered them invisible.

Toward those ends, and as a kind of epilogue to this chapter, MEJO released a report in the summer of 2008 based on its fish consumption surveys and focus-group results, outlining fish consumption disparities and recommending that lake-specific advisory signs, in Spanish, Hmong, and English, be permanently installed in most popular shoreline fishing locations (Powell and Powell 2008). The local public health agency in turn released its own report calling for increased outreach efforts and recommended that signs be installed in three languages at the three or four most popular shoreline spots. Although the agency report only adopted some of MEJO's recommendations, and framed the fish consumption and communication issues on its terms, it was a step forward.

The advisory sign project, however, encountered a surprising level of political resistance from natural resource agency leaders who did not want signs posted in city and county parks. MEJO continued to advocate for the signs—and for more than just three or four. In the fall of 2008, MEJO activists persuaded city and county elected officials to add just a few hundred dollars more to the advisory sign budget, which resulted in at least one sign in all of the most heavily fished urban locations. MEJO members designed the signs, making sure species that are eaten by many shore anglers but missing from advisories (e.g., white bass, catfish, carp) were included, and provided culturally appropriate Spanish and Hmong translations for the signs. The signs were subsequently approved by local and state public health agencies and the state natural resource department. Public health officials worked with city and county parks staff to place the signs where MEJO members recommended they be installed, based on their knowledge of the most popular shoreline fishing locations. It was a small—but hard-won—victory.

In spring 2009, MEJO worked pro bono with the public health department staff to develop a shoreline angler survey in three languages to evaluate the efficacy of the signs (one of the conditions placed on funding the pilot sign project by agency officials who resisted the project). In the summer and fall of 2009, MEJO helped train interviewers and conducted about 150 surveys with shoreline anglers in English, Hmong, Laotian, and Spanish. Public health staff did about 50 more interviews. MEJO's citizen scientist (Maria Powell) analyzed the quantitative and qualitative results and submitted them in a report to the public health agency in December 2009 (Powell, Xiong, and Powell 2009).

The survey results supported previous evidence we and others have gathered on consumption and awareness disparities (e.g., minority shore anglers eat significantly more fish than white anglers and are less aware of advisories)—but also provided useful information on where different kinds of anglers tend to get information about fish and what kinds of information they prefer. The signs are inspiring conversations and questions among shoreline anglers and others who spend time at the lakes about fish consumption risks, causes of water pollution, and ways to get more information and get involved. Most importantly, the interviews showed that shoreline anglers felt the signs were very useful for easily accessible and understandable fish consumption advice. Many anglers suggested that more signs be posted.

From this and prior experiences, our work has shown that while scientists and policymakers can be transformed situationally on specific issues, it takes ongoing diligence on activists' part to ensure that transformations are more than transitory and result in meaningful outcomes over time. In the projects we describe here, for example, we initiated the concerns, brought together the various local and state stakeholders, conducted the research, set forth plans of action, and advocated in public processes and via media over long periods of time to make sure they were carried out. This involved a tense "push and pull" between us and government agencies in which we had to work hard with few resources to initiate projects and make sure they were carried out. Public health professionals then used our work as "cover" to push decision makers to tackle environmental justice issues that would likely have fizzled out otherwise—or would not have been initiated at all. Of course, agency scientists' willingness to advocate on these matters helped us in turn. When decision makers listened and acted on our recommendations, it was because health agency leaders were willing to lend their credibility to our efforts. For now, though, MEJO has at least a tentative "place at the table" in community decisions about environmental health and justice. As to whether the evidence we gathered encourages public health or other government agencies to advocate for more signs or further work to address fish consumption risk disparities, as this book went to press the jury was still out.

Acknowledgments

We would like to acknowledge several University of Wisconsin students for their work and creative ideas in helping with several aspects of our projects. In particular, we would like to recognize Caitlin Dunn, Roxanne Felt, Cynthia Lin, Mary Parmer, and Ashley Viste.

Appendix I: Comparison of original fish consumption resolution (left) and substitute offered by county staff (right), and approved by the Dane County Board.

Note: The original language that was removed is struck through.

RES 238, 2007–2008

Posting of Fish Advisory Notices along Dane County Waters
~~Pollution in Wisconsin waterways has caused the State to issue fish advisory warnings regarding toxins to anglers and those who eat locally caught fish from inland Wisconsin waters. Levels of mercury, polychlorinated biphenyls (PCBs), polycyclic aromatic hydrocarbons (PAHs), pesticides, and other toxins are high enough in Dane County lake sediments and waters that people~~

SUBSTITUTE 1 TO RES 208, 2007–2008

Posting of Fish Advisory Notices along Dane County Waters
Fishing provides an enjoyable recreational opportunity for many Dane County residents, and fish from local waters are an important food source for many anglers. Certain species of fish contain contaminants at levels that pose potential health risks to people who eat fish frequently. The Wisconsin Department of Natural Resources, with

~~need to limit their consumption of fish~~
~~caught in these waters because these~~
~~compounds build up in fish tissue,~~
~~which humans consume. Yet, fish~~ advisory information is little known or unknown to ~~many~~ anglers.

~~Levels of mercury, PCBs and other~~
~~toxins that concentrate in fish are a~~
~~known public health hazard. Shoreline~~
~~anglers catch and consume many pan~~
~~fish that may have lower toxin levels~~
~~than larger fish, but when consumed in~~
~~high quantities they may exceed levels~~
~~recommended to avoid negative health~~
~~effects; they also catch and consume~~
~~larger fish, which tend to have higher~~
~~concentrations of toxins.~~

~~Public agencies have very little~~
~~actual data about local fish consump-~~
~~tion habits and toxin levels in locally~~
~~caught fish; and have little interaction~~
~~with local anglers and their families~~
~~who eat large amounts of locally~~
~~caught fish. Women of child-bearing~~
~~age, pregnant women and children are~~
~~especially at risk for developmental,~~
~~congenital and long-term risk from~~
~~exposure to toxins present in locally~~
~~caught fish. The environmental impacts~~
~~of pollution on low-income and minor-~~
~~ity citizens are often unknown or~~
~~underestimated because of a lack of~~
~~data collection, and lack of consider-~~
~~ation of these populations in determin-~~
~~ing public policy.~~

~~The common good and sound~~
~~public health policy is served by~~
~~informing anglers and others of poten-~~
~~tial risks associated with consuming~~
~~many kinds of locally caught fish.~~

advice and support from the Wisconsin Department of Health and Family Services—Division of Public Health, issues safe eating guidelines to anglers on size and species of fish to keep as well as how often and what quantities to eat, in order to reduce angler exposure to mercury, polychlorinated biphenyls (PCBs), and other contaminants.

For all waters in Wisconsin, WDNR advises that people follow these safe eating guidelines:

Women of childbearing years, nursing mothers and all children under 15 may eat: one meal per week of bluegill, sunfish, crappies, yellow perch, bullheads and inland trout; AND one meal per month of walleye, pike, bass, catfish, and all other species. Do not eat muskies.

Men and women beyond their childbearing years may eat: unlimited amounts of bluegill, sunfish, crappies, yellow perch, bullheads and inland trout; AND one meal per week of walleye, pike, bass, catfish, and all other species; AND one meal per month of muskies.

For Badfish Creek, Lake Mendota, Lake Monona, and the Wisconsin River in Dane County, WDNR also advises that in order to further reduce exposure to PCBs, people should eat no more than one meal per month of carp. For the Wisconsin River, people should also eat no more than one meal per month of Lake Sturgeon. PCBs are generally stored in the fat of fish, so people are advised to reduce PCB levels in fish they eat by trimming

~~NOW, THEREFORE BE IT RESOLVED, that the Dane County Board of Supervisors directs the Dane County Parks Department to post and maintain fish advisory notices at all boat landings and other county owned land where shoreline anglers fish, advising anglers of potential risks associated with consuming locally caught fish. Said notices shall be in English, Spanish, and Hmong in non-technical language understandable to the average person, and shall be posted within 90 days of the effective date of this Resolution.~~

BE IT FURTHER RESOLVED that the Dane County Board of Supervisors requests that the Department of Public Health for Madison and Dane County ~~address this issue and~~ work with the Dane County Lakes and Watershed Commission, ~~and~~ appropriate local and state agencies, ~~as well as~~ environmental justice organizations and affected communities ~~of color, to post and maintain fish advisory notices along all Dane County lakes and waterways.~~

away the fatty areas and removing the skin before properly cooking their fish.

This fish consumption advisory information is little known or unknown to some anglers.

NOW, THEREFORE BE IT RESOLVED that the Dane County Board of Supervisors requests that the Department of Public Health for the City of Madison and Dane County investigate existing outreach efforts advising anglers in English, Spanish and Hmong and in non-technical language understandable to the average person, of potential health risks associated with consuming locally caught fish.

BE IT FURTHER RESOLVED that the Dane County Board of Supervisors requests that the Department of Public Health for the City of Madison and Dane County work with the Dane County Lakes and Watershed Commission, appropriate local and state agencies, environmental justice organizations and affected communities to recommend the most effective mechanisms for educating local residents about potential health risks associated with consuming locally caught fish. These mechanisms may include signage, educational events, and small group meetings.

7

Risk Assessment and Native Americans at the Cultural Crossroads: Making Better Science or Redefining Health?

Jaclyn R. Johnson and Darren J. Ranco

Environmental justice (EJ) activists and communities have offered many valuable critiques of environmental scientific management over the last thirty years. In the realm of environmental risk assessment, these critiques have led to some new practices and uses for science and engineering in the service of EJ concerns, such as mapping the unequal distribution and impact of environmental "harms" in our society (see Harris and Harper 1997; Allen 2003). These new uses for environmental science go hand and hand with an emerging science community with origins in EJ communities as well as science activists seeking out a worthwhile cause for their work (see Frickel, chapter 1, and Nieusma, chapter 5, this volume).

Many critics of environmental risk assessment have pointed out that the science used in risk assessments does not protect cultural minorities very well for a variety of reasons—one of the key reasons being that some cultural minorities access resources in the environment differently than both the mainstream culture and the scientists and policymakers doing the risk assessments (Kuehn 1996; O'Neill 2000; Powell and Powell, chapter 6, this volume). Since the late 1990s, there has been a flurry of activity within both the Environmental Protection Agency (EPA) and American Indian Nations to address this problem, and two different but sometimes overlapping approaches have developed to address the problems of cross-cultural risk assessment. This chapter is an exploration of how the EPA and Tribal Nations have shaped these approaches.

One approach to solve the problems of risk assessment that the EPA and Tribal Nations have explored is to make science more responsive to the ways that Native Americans actually access resources in the environment both now and in the past. Instead of using aggregate models from the entire American population for something like fish consumption, this approach would have the EPA and tribal scientists measure the actual intake of fish from tribal people living off of the resources either currently or historically. The other approach is much more culturally relevant but harder to define in a scientific manner from the EPA's perspective. This approach, defined by some as the "Health and Well-Being" model, would allow for and encourage Tribal Nations and the EPA to redefine health itself in culturally relevant terms. In this model, a risk assessment would not just emphasize the potential deaths to a population caused by cancer, but would look at the risks to a healthy, culturally defined lifestyle in each Tribal Nation. This approach would potentially allow Tribal Nations to redefine health in much broader terms than cancer death rates and would include cultural indicators such as access to a healthy traditional diet, ability to participate in ceremonies, the passing down of traditional knowledge, and so on.

In this chapter, we will examine these two approaches to fixing environmental risk assessment, examining how Tribal Nations engaged the EPA and its risk-assessment procedures, while at the same time trying to create their own culturally appropriate risk assessments. The case is a clear example of how EJ communities have begun to transform environmental science. This particular transformation is far from complete, however, which is itself revealing. Any challenge to normative approaches in science will face serious opposition; in this case, it comes from the administrative authority of the EPA and its power, as an agent of the state, to shape normative processes, especially the participation of Tribal Nations and other "lay" groups in policymaking. As a key theme of this book, then, we should understand that EJ transformations of science are almost always tied to calls to transform participation, and this may be the more challenging project (see Cole and Foster 2001).

We should also point out that the normalizing qualities of traditional risk assessment likely fail to protect non–Native Americans as well as Native Americans—anyone who lives in a nonsuburban lifestyle is probably being impacted by pollution unfairly. Thus, while we look at a series of engagements between the EPA and Tribal Nations in this chapter, the general themes apply to any EJ group and their allies trying to get new forms of knowledge sanctioned by the state. Moreover, the call by some tribes and tribal scientists to shift regulatory decisions away from risk assessment and risk management to a more complete picture of environmental hazard and human health are themes in other chapters in this book (see Powell and Powell, chapter 6; Liévanos, London, and Sze, chapter 8). While many critics can point out that the system is broken, it takes a huge amount of effort to change what counts for knowledge in a regulatory context.

To address these issues of power and knowledge, we organize our chapter in the following way. After looking carefully at the critiques around environmental risk assessment, we look at how the EPA and Tribal Nations started to work on the problems of risk assessment together through the national EPA's Tribal Science Council (TSC), essentially blurring the lines between tribes and the EPA. We also look at the different forms of risk assessment being offered by groups of scientists working from and within Tribal Nations and how their "new" science impacted the TSC and its work on environmental risk assessment. From there, we assess the different approaches brought to the TSC from Tribal Nations and how EPA scientists and managers responded to them. Based on our own experiences working at the EPA and with Tribal Nations, we conclude with thoughts on why the TSC engagement has done little to impact EPA risk-assessment protocols.

Environmental Risk Assessment and Its Critics

A number of scholars, most notably Robert Kuehn and Catherine O'Neill, have pointed out that the current ways that scientific managers assess the health risks posed by pollution are flawed, especially with respect to the risks faced by cultural and racial minorities (Kuehn 1996; O'Neill 2000). They each examine how the four stages of the scientific risk-assessment process identified by the National Research Council in 1983—hazard identification, toxicity or dose-response assessment, exposure assessment, and risk characterization (National Research Council 1983, 3)—have biases that expose racial and cultural minorities to much more risk than other populations because of the ways in which the science is done, and the kinds of assumptions it makes about scientific management.

In the hazard-identification stage, regulators have to assess the overall toxicity of a substance. Because not every chemical must be tested before it is put into the environment, there are gaps in the knowledge, and regulators are often forced to use chemical proxies for the chemical being regulated, relying on the similarity of elemental structure of the chemicals to justify the comparison to other, more tested substances (O'Neill 2000, 28). This may be okay if one is only being exposed to a few chemicals, but the more chemicals a person is exposed to, the more likely that a proxy will underestimate the toxicity of a particular chemical agent. As a result, the use of proxies has the potential to cause more damage to highly exposed communities.

During the dose-response assessment, where a regulator must decide what health impacts are posed by a certain amount of a particular chemical, there are also gaps in knowledge that may harm EJ communities. Because there is an absence of human epidemiological data, regulators must use and extrapolate from animal testing (O'Neill 2000, 28). The use of animal data can introduce significant uncertainty to dose-response assessment; for example, hamsters are several thousand times more resistant to dioxin than guinea pigs in lab tests (Steingraber 1997, 223). When such a disparity in testing exists, regulators typically choose to treat human sensitivity to such chemicals somewhere between the two extremes. In communities that are exposed to many chemicals, synergistic effects often go untested; thus, there is an increased chance of making a decision that harms the health of those communities.

The exposure assessment phase of risk assessment looks to understand the fate of chemicals in the environment and the potential exposure pathways of chemicals into human beings. Many commentators have pointed out that risk-assessment models grossly underestimate these pathways for cultural and racial minorities, especially those living closer to the environment in "lifeways" or lifestyles that use subsistence resources in the environment for food as well as other cultural activities, such as the gathering of grasses for basket-making or using river water for ceremonial practices. Stuart Harris and Barbara Harper (1997, 794) have shown that Native American lifeways, both through subsistence and other cultural practices, result in an exposure to pollutants that is two to a hundred times greater than the standard exposure models used by risk assessors. Also, risk assessors do not typically account for the synergistic effects of chemicals, which would be assessed in this phase if there were a mandate to understand how chemicals interact with one another and create more harm in human beings (O'Neill 2000, 28).

In the final stage, risk characterization, environmental managers must "produce an estimate of the type and magnitude of the effect that will occur from exposure and probability that each effect will occur" (Kuehn 1996, 114). The decision is a determination of how much of a particular chemical is allowed to discharge into, or remain in, the environment. Thus, based on a series of scientific uncertainties, environmental managers must still decide how to regulate a substance, and many critics cite this as a key part of the problem—the more uncertainty, the more likely overexposed populations will bear a greater and unfair amount of risk

(O'Neill 2000, 29). Kuehn (1996, 131) is particularly concerned that in each stage of this uncertainty, scientists, who do not live in the most affected communities, make decisions about people's lifestyles and behaviors that are ethnocentric or disconnected from the reality of those impacted the most by pollution. In fact, Kuehn is worried that the risk-assessment process "reduces complex social problems to a number" (p. 126) and that it "transforms disputes over values and politics into scientific disputes that are inaccessible to many citizens and may be particularly inaccessible for communities of color" (p. 171).

In the typical risk-assessment process, "stakeholders" are often included in the last stage. They are allowed to comment on the logic that drives the management of risk, but not on the science that drives these decisions. To fix this problem, Kuehn calls for "risk bearers to be made meaningful participants at the beginning of the risk assessment process" (p. 161). Tribes and tribal researchers have been at the forefront of this kind of criticism (see Harris and Harper 1997; Arquette et al. 2002), and with the creation of the Tribal Science Council in 1999, they were able to engage the EPA in a more direct manner, with the goal of changing environmental risk assessment.

The fact that the EPA has created the TSC at all reveals how the EPA views the public's involvement in science (or at least tribal involvement in EPA science). Thus, the story of the TSC is not just about which numbers should be used when Native Americans are being impacted by pollution, but also how science is conceived and used in the regulatory processes. Sheila Jasanoff (1992) documents the changing ways in which the EPA divided the lines between the "public" and "science" through the 1970s and 1980s—insights that are instructive here. Over this period, she notes that the EPA's "public representation of science shifted away from an emphasis on testable knowledge claims to a preoccupation with the processes of knowledge production" (p. 197). On its surface, this shift could be understood as a positive shift toward an inclusive pluralism that many EJ scholars advocate. For example, Eileen Gauna (1998, 57) has called for an emphasis on process that opens up more "participation avenues" for EJ groups.

While this pluralist self-reflection over the process of knowledge production can be seen as a positive thing for environmental groups and Tribal Nations in the case of the TSC, Jasanoff is critical of such an approach. Why? One reason to worry about this emphasis on process seems to be the potential challenge to EPA expertise by a lay public—to her, the blurring of lines creates more problems than it solves. The emphasis on process, she argues, has "created an almost intolerable dissonance between the actual practice of policy-relevant science, which tends to assimilate science to politics, and the public's continuing demand for decisions based on politically untainted knowledge" (Jasanoff 1992, 197). Despite this dissonance, however, the emphasis on process has also allowed experiences and knowledge to make their way into the EPA that would not normally be there in the realm of "science" through avenues like the TSC. We will explore these themes in more detail later in the chapter when we examine the impacts of the TSC on environmental risk assessment.

The Tribal Science Council

Throughout the 1990s, Tribal Nations voiced their concerns about the EPA's policies and scientific practices. This timing is consistent with Jasanoff's observation about the EPA's shift from knowledge producer to moderator of the process of knowledge production and policy formation—in many ways, during the 1990s, the EPA was open to, and seeking input from, a variety of groups including EJ activists and industry lobbyists. During this time, Tribal Nations expressed concerns about the "appropriate use of scientific information gathered by tribes, the validity of data collected, and the ability of the EPA's programs to incorporate the unique aspects of tribal cultures into its models and assessment tools" (Tribal Science Council 2002, 2).

In May 1999, Norine Noonan, former Assistant Administrator of the Office of Research and Development in the EPA, called for the formation of a Tribal Science Council to provide a structure to tribal involvement in the EPA's science efforts (Tribal Science Council 2002, 2). The TSC's first meeting was held in December of 2001 (Tribal Science Council 2002, 2).

The TSC consists of both tribal and EPA representatives and operates under the auspices of the Office of Research and Development (ORD) at EPA headquarters. It is separate from, but has collaboration with, the American Indian Environmental Office in the Office of Water, but derives its funding from ORD. One tribal representative is included from each of the nine EPA regions that have federally recognized tribes, including a representative from Region 10 to represent Alaska Native communities, and is responsible for representing all tribes within the specific region. Tribal representatives, who are selected by the Regional Tribal Operations Committee (R-TOC), "serve as a liaison for tribes within their region to identify tribal science priorities and implement the mechanisms identified for addressing these priorities" (Tribal Science Council 2002, 4). There is also an EPA representative from each Headquarters Program Office and Region. EPA representatives' responsibilities include: "representing their specific office or region, authorization to assess tribal science priorities relative to their region's ongoing activities, and serving as a liaison with other tribal program activities supported by their office" (Tribal Science Council 2002, 4). The TSC consists of two cochairs that serve two-year terms. One chair is an EPA representative and the other is a tribal representative. The cochairs' roles are to coordinate meetings and discussions, provide guidance, and speak on behalf of the TSC (Tribal Science Council 2002, 6). The TSC also has an Executive Secretary, selected by the Office of Science Policy within the ORD, to "help ensure the effective operation of the Council and facilitate the flow of information within and beyond the Council" (Tribal Science Council 2002, 7).

The creation of the TSC's mission statement was a combined effort that involved both EPA and tribal representatives. The statement reads: "The mission of the Tribal Science Council is to provide a forum for interaction between Tribal and Agency representatives of mutual benefit and . . . to support the subsistence, cultural, and ceremonial lifestyles of Indians and the safe use and availability of a healthy environment for present and future generations" (Tribal Science Council 2002, 3). The TSC's goal is to "develop a better understanding of the priority science issues of Tribal Nations and the EPA's ability to address these issues as part of its formal planning process" (Tribal Science Council 2003b). The TSC works to form a stronger partnership between tribal and EPA scientists and facilitate the communication and coordination with other agencies to more effectively respond to these issues (Tribal Science Council 2003b).

Ideally, the TSC is driven by tribal priorities. The tribal representatives identify concerns from their region, and then together as a group they prioritize the concerns by creating a list of priorities, which are then discussed by the TSC (Tribal Science Council 2002, 8). Thus, the TSC serves as a forum for tribes to raise tribal concerns, directly from the voices of Tribal Nations and their representatives. TSC members coordinate with tribes, tribal scientists, tribal organizations, and tribal colleges as an integral part of their work (Tribal Science Council 2002, 4). As Dan Kusnierz, a TSC representative from the Penobscot Indian Nation in Maine, stated, "TSC has been an opportunity for tribes to get their science issues on the radar screen" (Kusnierz, phone interview with Johnson, March 7, 2005).

In September 2002, the TSC identified eight tribal science priorities that the TSC would address to help guide its future activities. They included traditional tribal lifeways, including risk assessment; endocrine disruptors; dioxin reassessment and reference dose; cumulative impacts; persistent organic impacts; toxic molds; pharmaceuticals in wastewater; and tribal research (Tribal Science Council 2003). Following the creation of this list, the TSC conducted a number of different workshops and presentations dealing with these priorities. Noteworthy events include an Endocrine Disruptors Workshop in 2002, a Risk Assessment and Health and Well Being Workshop in February 2003, and a Health and Well Being Workshop in May 2003 (Tribal Science Council 2003a, 2003b). The TSC activities have worked to expand tribal participation and tribal scientists' involvement in EPA activities. Dennis O'Connor, former TSC representative from the Office of Air and Radiation, stated that the accomplishments of the TSC could be found within the narrow, specific topics. The TSC has been successful in working with "small technical problems, while larger issues become much more difficult for the voluntary council who has to take time out of their regular work" (O'Connor, phone interview with Johnson, February 11, 2005).

The TSC Tackles Risk Assessment

When the TSC identified its tribal science priorities, the EPA's risk-assessment approach made the list as a part of the priority labeled "traditional tribal lifeways." In an interview, the TSC member from the Penobscot Nation mentioned earlier—Dan Kusnierz—highlighted the importance of risk assessment, saying he feels "it is the nexus of where so many tribal problems merge" (Kusnierz, phone interview with Johnson, March 7, 2005). The Health and Well Being Workshops, referenced above, were not the first time the risk-assessment discussion had surfaced in tribal-EPA relations. Before the TSC's inception, a conference dealing with risk assessment was held with tribal and EPA representatives, and after the creation of the TSC, tribal representatives stated that they "felt that not enough was happening in the area of risk assessment" (Tribal Science Council 2003b, 5). There was a strong desire to continue these discussions and the TSC was clearly the forum for doing so. Tribal representatives felt that because the TSC was a creation of the EPA, this would be an effective place to influence policy changes.

As discussed earlier, the EPA's risk-assessment process is not equipped to accurately assess and confront the risks facing tribal communities and specific tribal needs—the process used by the EPA essentially ignores tribal cultural differences (and really any other kind of difference that deviates from the suburban lifeway). It is much easier for the EPA to apply one already established method used for all Americans, rather than to create a different process each time a tribal culture is impacted by a source of pollution.

In the 1990s, before the creation of the TSC, most tribal attention focused on risk assessment that had to do with fish consumption rates—the EPA had been unwilling to set a consumption rate for subsistence users of water resources (see Ranco 2000; O'Neill 2000). The consumption rate has a direct relationship with the third phase of risk assessment, the exposure assessment phase, because establishing an ingestion rate of potentially contaminated fish is one of the few ways the standard risk-assessment model tracks pollution into human beings. At the urging of the Columbia River Inter-Tribal Fish Commission and other advocates and scientists, in 1997 the EPA finally set a subsistence consumption rate of fish for Native Americans at 70 grams per day—three and a half times the mean ingestion rate of the overall U.S. population (U.S. Environmental Protection Agency 2000a).

The fact that the EPA was starting to look at, to a certain extent, the actual consumption rates of Native people was only a partial victory leading up to the creation of the TSC. Researchers from a number of Native communities across the country were starting to call for risk-assessment models that included all the potential exposure pathways in tribal communities. These models go far beyond looking at fish consumption rates and include exposures to pollution in all aspects of a subsistence lifestyle, including, but not limited to, potential exposures during the gathering of plants for medicinal and other uses, higher water and air ingestion rates during hunting and fishing, and the exposures to soil, air, and water in other cultural activities like sweat lodges (see Harris and Harper 1997; Arquette et al. 2002; Harris and Harper 2000; Powell and Powell, chapter 6, this volume).

Better Science by Redefining Health

Reforming the science of environmental risk assessment to measure all these pathways of exposure is one approach that tribes are taking to fix risk assessment to meet their cultural and health needs. This approach of making better science is consistent with Kuehn's call to expand risk assessment to provide a complete characterization of the risks that minority and low-income populations face (Kuehn 1996, 21). It represents the production of new scientific processes instigated by the search to promote environmental justice for those populations. Many of the problems of risk assessment can be dealt with by conducting science that engages Native Americans and their exposure scenarios.

Stuart Harris and Barbara Harper at the Confederated Tribes of the Umatilla Indian Reservation (CTUIR) head one group that has done this kind of work, looking at the actual exposure scenarios of traditional lifestyles. Working with tribal members at CTUIR and other tribes, they have started to develop a lifestyle-based subsistence exposure scenario for Native subsistence users (Harris and Harper 1997, 2000). Using in-depth interviews and other ecological information, they have evaluated risks by including "exposure factors for subsistence activities and diets as well as factors for environmental and socio-cultural quality of life" (Harris and Harper 1997, 789).

Another group of researchers, from the Akwesasne Task Force on the Environment and the Haudenosaunee Task Force on the Environment, working primarily with and from within the Akwesasne Mohawk Nation, started to address these potential pollution pathways in subsistence diets during the 1990s as well. Led by Mary Arquette, Katsi Cook, Brenda LaFrance, Jim Ransom, Arlene Stairs, and others, this group, much like Harris and Harper, called for "better site- and Nation-specific data" for true risk assessment to take place (Arquette et al. 2002, 261). They called for this information to inform an indigenous decision-making process "regarding the effect of contaminants on health" (Arquette et al. 2002. 261). They were particularly concerned that, if the health effects of pollution were finally known, outside policymakers would encourage risk-avoidant behavior in the tribal population, and that this, in turn, would lead to other detrimental effects to people's health and culture by discouraging them from engaging in subsistence lifeways. For the Akwesasne Task Force on the Environment, the indigenous decision-making model, which includes "the use of traditional values and political systems" (Arquette et al. 2002, 261), counters this kind of risk-avoidant management and is an expression of tribal sovereignty.

While all of these groups of researchers have called for more scientific study of the actual environmental risks faced by Native American populations, they have also called for a change in the ways risk assessment is conceived. Calling for alternatives to classic risk assessment, Arquette et al. (2002, 262) call for a new paradigm "that not only recognizes the requirement for unique and shared decision making with Native governments but that also recognizes the important role that community-based research, specialized communication strategies, and community participation play in decision making." This process-oriented approach to doing risk assessment is supported by culturally rooted ideas about tribal sovereignty and definitions of health—as Arquette and colleagues put it, "Native people need to have opportunities to meet their own physical, mental, emotional, spiritual, social, and ecological needs using their own culturally defined paradigms" (p. 262). Harris and Harper (2000, 92) have also pursued a more culturally defined notion of human health in their work, calling for "new integrating tools" between the impacts of pollution on culture and human health. For them, "evaluating impacts to a traditional way of life would include environmental and community quality of life in addition to personal exposures to contaminants" (Harris and Harper 1997, 793). These kinds of impacts go far beyond the classic risk-assessment model and would allow for tribal communities to define health in culturally appropriate ways, not limited to the cancer death rates caused by toxic pollutants.

Picking up on this research, the TSC proposed that a new paradigm be created. While early discussions focused on the need to improve the technical aspects of risk assessment, it is this second approach, named the Health and Well-Being paradigm, that the TSC has pursued more actively and tribal representatives have advocated most urgently. It is the TSC's vision to have this new paradigm "incorporate the cultural (spiritual, emotional, physical, and mental) interconnectedness between tribes and the natural world into the assessment" (Tribal Science Council 2002, 1).

The Health and Well-Being Paradigm

The Health and Well-Being paradigm emerged in September 2002 from TSC discussions about risk assessment, these discussions originally being a component of the tribal science priority list and workshop on traditional tribal lifeways. In this workshop, the TSC started to recognize that risk assessment is not suitable for tribal communities: "The current risk assessment and management methodologies do not take into account Tribal culture, values, and traditional lifeways" (Tribal Science Council 2003b). Concerns about how risk assessment ignored these important factors were heard throughout the TSC. Jeannette Wolfley, a tribal representative from the Shoshone-Bannock tribe, noted at a TSC workshop in February 2003, "While cultural factors, such as the impact of a potential action on a Tribe's origin or creation story, landscapes, historical stories, songs, dances, prayers, language, etc., may not be easily quantifiable and are ignored by current risk assessment frameworks, these factors are vitally important to the continued health and well being of Tribal communities" (Tribal Science Council 2003b, 6). Different TSC workshops reiterated similar concerns and opinions from tribal representatives; thus, discussions progressed on creating the alternative Health and Well-Being paradigm.

According to the TSC, the Health and Well-Being approach seeks to create a process that is completely tribally driven. A tribal approach would serve as an alternative to the current risk-assessment approach and allow full tribal participation from the beginning. Many advocates have viewed the idea of a tribally driven approach as a long-term goal, while the short-term goal is to amend the current risk-assessment approach (Kusnierz, phone interview, March 7, 2005). Under the Health and Well-Being approach, tribes would create their own definition of a "healthy" community and then develop standards and manage a plan that would maintain this healthy environment.

The TSC recognizes that the EPA's risk assessment applies a generic definition of "health" to tribal communities and that tribes often possess different definitions of a healthy community. The risk-assessment process identifies risk by using indicators that are "based on measurements of illness, dying and death" (Tribal Science Council 2002, 3). In contrast, the "health" of tribal communities can have many definitions that are centered on quality of life. For the Mohawk people of Akwesasne, health is rooted in the culture; it is based on "peaceful, sustainable relationships with other peoples including family, community, Nation, the natural world, and spiritual beings" (Arquette et al. 2002, 262). The EPA's risk assessment does not recognize the different tribal views of "health."

The new paradigm has been promoted by a core group of tribal representatives. One of the prominent advocates and spokesmen has been Jim Ransom, member of the St. Regis Mohawk Tribe, who documented the initial Health and Well-Being concepts and presented them to the TSC. TSC Executive Secretary Claudia Walters played a large role in organizing the workshops discussing the new paradigm. A number of tribal representatives have participated in these workshops, but Walters also brought together a smaller group to further progress with the paradigm and to create a Health and Well-Being concept paper. She has identified these people as strong tribal representatives who have worked on efforts similar to those that will be needed for the new paradigm; thus, they will be effective participants in furthering the Health and Well-Being paradigm. These tribal representatives include Terry Williams from the Tulalip Tribe in Washington State, Larry Campbell from the Swinom-ish Tribe in Washington State, Jeanette Wolfley from the Shoshone Bannock Tribe in Idaho, and Brenda LaFrance from the St. Regis Mohawk Tribe in New York. Conference calls, emails and meetings occurred among Walters and this group between 2002 and 2006, and they have continued to present their work to the TSC as a whole.

Workshops and seminars have become the main forum for discussing the Health and Well-Being paradigm. Over the past several years, a few TSC workshops and meetings have been held to discuss and move forward with the new approach. It is in the early stages, as members of the TSC discuss how to define the paradigm and construct a working model. In May 2003, TSC held a workshop titled "Traditional Tribal Lifeways: Health and Well-being Workshop." One of the goals was to explain the Health and Well-Being paradigm to individuals outside of the TSC, specifically to tribes and EPA representatives not currently involved with the paradigm. The TSC explained how the new paradigm would be an alternative approach to risk assessment. In the words of Jim Ransom, "The Health and Well-being Paradigm is an attempt to take a more positive approach to assessing the health of the world around us. If we are focused on a healthy environment, we should focus on what it takes to make the world healthy and make ourselves (Native Americans) healthy" (Tribal Science Council 2003a). In February 2004, the TSC convened once again on the Tulalip Reservation for the "Health and Well Being Paradigm Development Meeting." One focus of the meeting was to review the paradigm and how it differed from the current risk-assess-ment paradigm. At this meeting the following Health and Well-Being paradigm concepts were documented:

• Driven by tribal priorities that are identified by tribes.

• Ways to identify these tribal priorities include the cultural ecosystem stories approach, resource audits, interviews with tribal elders and similar processes.

• Priorities emerge from a discussion of the past, present, and future for a tribal community.

• Focuses on cultural landscape management.

• Is holistic in nature, encompassing all aspects of tribal community relationships.

- Addresses priorities beyond environmental concerns.
- Incorporates impacts that occur on and off tribal lands.
- Promotes development of creative approaches for filling gaps.
- Will involve multiple Federal agencies and other levels of government.
- Closely relates knowledge and resources, which cannot be separated.
- Is based on the premise that for tribal people to be healthy, it requires healthy environment and healthy culture.
- Personal experiences, observations, and historical knowledge are backed up and are supported by empirical scientific data. (Tribal Science Council 2004, 9)

Unlike the classic risk-assessment paradigm, the new Health and Well-Being paradigm addresses tribal cultural differences. It is driven by tribal priorities, whereas risk assessment is driven by the EPA's regulations and measurements of risk (Tribal Science Council 2004, 9). It also focuses on the health of communities, which is defined by the tribes. A healthy community encompasses all aspects of tribal relationships and tribal priorities that affect a community (Tribal Science Council 2004, 9). The paradigm does not simply focus on a certain aspect of the environment; it will take a holistic approach so that the interconnectedness of all aspects of a community is respected.

To build a framework for this new paradigm, tribal input is vital. Tribal input is needed to determine what a healthy community is and to articulate tribal culture. As the meetings progressed, it was understood that persons outside of tribal communities, on their own, are unable to do this. The TSC acknowledges the importance of meaningful tribal participation and knows tribal participation is needed in order for the new paradigm to be tribally driven and in order for the problems of risk assessment to be confronted and remedied. As Dan Kusnierz stated, "Engaging the tribal communities and involving them early on in the process is a big part of it" (Kusnierz, phone interview with Johnson, March 7, 2005).

Pluralism and the Problem of Recognition

As we can see, the Health and Well-Being paradigm not only includes different forms of knowledge in understanding environmental risk; it also calls for a different process of science production when protecting the cultures of Tribal Nations. As we suggested earlier, attempts at incorporating tribal concerns into EPA science can blur the lines between policy, science, and activism, and, as articulated by Jasanoff (1992), this can have serious negative consequences for an administrative agency like the EPA. For one, the blurring opens up EPA science to any subgroup in the United States and thus weakens its efficacy in the public domain, *to this same American public*: "EPA was committed by law and cultural tradition to making rational decisions, and for the American public rationality meant that science should be kept distinct from politics and policy" (Jasanoff 1992, 207). Jasanoff is concerned that this pluralist discourse is too much of a challenge for the science used by agencies like the EPA. While she admits that science and democratic politics have mutually reinforced one another with their ideas of open challenge and transparency, she believes that "when science itself is part of the matter that must be made transparent . . . [it] becomes indistinguishable from the political goals to which it is harnessed" (pp. 216–217). In fact, this critique joins with others by noting that the EPA does not and cannot do relevant science, not because they are an arm of the government and therefore are political, but because their scientific practices are made political by the legal process engendered by pluralist discourses. The openness required by judicial review and active public involvement "has reversed the ordinary processes of scientific fact building, leaving EPA without a credible foundation for regulatory action" (Jasanoff 1992, 217).

Our research supports the view that since the 1970s, the EPA has become less of an "expert" agency and more of a mediator of various political and scientific interests. For EJ groups, especially Native Americans, this certainly cuts both ways—there is greater access and possibility for comment and involvement, but it is also more difficult to effect a change since other, more powerful, political interests have similar access and may undermine EJ interventions. Moreover, as "minority" concerns in a political (not scientific) scene, it might seem that this turn toward pluralism would make it even more unlikely to change the uses of science by regulatory agencies in your favor. This provides us with an important mapping of the power in such situations and of how we might want to understand the TSC. Whereas in the early history of the EPA, the agency could overlook tribal concerns because they were tasked with developing a general science to serve the public (which, as we pointed out earlier, had certain cultural and experiential biases), we now have an agency that exercises its power as a mediator of interests and exercises its power in a different way. It still, though, has the power to decide who and what get included in such a political process.

To amplify this point about forms of power, consider the recent history of the TSC and its interest in risk assessment since the flurry of meetings between 2001 and 2004. During coauthor Johnson's internship at the EPA in the summer of 2004, the TSC was trying to create a Health and Well-Being concept paper. The concept paper's purpose was to outline components such as the needs of Indian Country, tribal priorities, the guiding principles of the new paradigm, and management issues. Johnson was to outline a tribal model for the new paradigm, and identify what the next steps to incorporating this model would be. And after a model had been formed, the paper was supposed to be presented to the EPA and the TSC for review. After this review, selected tribes were to act as demonstration pilots for the paradigm (Tribal Science Council 2004, 7). The TSC had been looking for input from tribal representatives on developing this model for some time before Johnson began her internship (Tribal Science Council 2003a, 12). At the TSC Workshop in 2003, Jim Ransom discussed the "need to develop a common vision of tribal health and well-being. While differences exist among all tribal communities, it is necessary to identify key components of health and well-being and build from these" (Tribal Science Council 2003a, 12). Tribal input, which was to form the basis of Johnson's report, was to be vital in identifying these key components.

Johnson's work focused on framing this general model of the Health and Well-Being paradigm. Claudia Walters, the TSC Executive Secretary, briefly described the idea of the Health and Well-Being paradigm to Johnson, but explained that no explicit model had been formed. Johnson's job was to research possible examples of this paradigm: Were there tribes that already implemented the paradigm's concepts or exhibited features similar to the concepts? Then, by looking at these examples, a framework for a model could develop. Referring to notes from the first day, three central terms came to the fore in Johnson's search: traditional tribal lifeways, culture, and holistic approach. She was to frame her search for Health and Well-Being models around these three terms.

These three terms became essential in Johnson's search and crucial for creating a model. A tribe had to use at least one of these words in their official policy documents in some way in order to be included in the framework of the model. This is an example of how an administrative agency controls participatory approaches to science such as this. In this context, the EPA has the authority to decide what will and will not work for a tribal model. Johnson quickly found that some tribes "fit" the terms and description the EPA was looking for, while others did not. An example of a tribal model that did not fit was Johnson's own tribe, the Confederated Salish and Kootenai Tribes (CS&KT) on the Flathead Reservation. The CS&KT have established an Environment and Natural Resources Department guided by this mission statement: "To advocate for the protection, restoration, effective management and health of the air, land, water, and biological resources of the reservation environment through education, planning, conservation, cooperation, and regulation, while perpetuating cultural values and the quality of life of the people of the Flathead Indian Reservation" (Confederated Salish and Kootenai Tribes). From reading this mission statement, the CS&KT seemingly fit into the concepts of the Health and Well-Being paradigm and could prove to be helpful in forming a model. But the CS&KT program was not seen as an example of Health and Well-Being by Walters and others at the EPA. The CS&KT work on separate environmental issues that are of high priority, but do not necessarily utilize a "holistic approach" in addressing them. The new paradigm seeks to take a holistic approach toward environmental and community issues. Furthermore, Confederate Salish and Kootenai tribal "culture" or "lifeways" are not specifically stated as being used and incorporated in each program, and this was also not considered to fit into the Health and Well-Being model. Johnson encountered this pattern repeatedly, observing more than one case where a tribe had a certain approach to their environmental issues but did not explicitly say they were integrating their tradition or culture, or where a tribe focused on a certain environmental issue instead of taking a "holistic approach" by incorporating all environmental issues. In these cases, EPA officials insisted that they could not be used as examples to help form a basis for the Health and Well-Being paradigm.

On the other side, the ORD and EPA saw tribes that explicitly mandate the incorporation of their culture and a holistic view in addressing community health as prime examples for a tribal model. The Akwesasne Task Force on the Environment has become a leader in this respect in the eyes of the EPA, and over the years it has had strong representation on the TSC. On websites, in programs, and in documents, the Akwesasne use phrases or words like "traditional lifeways," "culture," and "holistic approach." They also have strong representatives and spokespeople, a model of engagement that the EPA favors. In this process of forming a tribal model, the EPA has taken control of how to define the tribal model. Decision-making power is left in the hands of EPA bureaucrats who rely on their day-to-day forms of recognition of Native concepts. The tribal model is specifically an EPA image of what the Health and Well-Being paradigm should be; however, this is supposed to be a tribally driven paradigm.

Returning to the points about the historical shift in the EPA's approach to risk assessment, we can see some critical themes about the exercise of administrative power in situations where the EPA is balancing interests and not finding facts. Control over the process manages facts and citizens in very specific ways, although it may not always be clear what they are. In political theory, the kind of power that manages citizens in this way is referred to as the "politics of recognition" (Taylor 1994). This is an issue that is particularly salient to cultural minorities in liberal states, especially indigenous people, because the universal "difference-blind" politics often "negates identity by forcing people into a homogeneous mold that is untrue to them" (Taylor 1994, 43). In the case of quantitative risk assessment and Native Americans, the forms of recognition have functioned in two ways. The first way is more typical—the science that the EPA uses to assess environmental risk does not protect the interest of any particular subgroup or cultural minority, but instead imposes a general model of citizen protection. The second way is more difficult to ascertain, because the state, in the guise of the EPA, is managing the process and categories of political recognition—this can be found in the problems associated with Johnson's report and in the form of the TSC itself.

This second type of the politics of recognition is brilliantly detailed in the work of Elizabeth Povinelli in her scholarship on Australian aborigines (Povinelli 1997, 1998, 2002). She points out that this newer form of liberal state management, where the cultural other is included in the state's apparatus, firms up the power of the state in even more dramatic ways than previously. In these instances, in her view, this form of partial recognition functions as a "double gesture." On the one hand it "makes visible a particular set of social formations and subjectivities which necessarily have, more or less, a relationship to social reality" in any particular historical or geographic moment. On the other hand, it "constructs the interiors and edges of legitimacy—the boundary past which, say, the 'special rights' of native title turn into the 'equal rights' that the state provides some minorities" (Povinelli 1997, 20). That is, for Povinelli, liberal states, as the arbiters and generators of recognition in such cases, will always refashion calls for "special rights" in a way that is consistent with the liberal project of "equal rights" and will therefore erase and fail to fully recognize and endorse the call for special rights brought by indigenous people in these instances. The problems that Johnson encountered included a superficial, homogeneous, understanding of the category of "Native," an obstacle that is all too typical in such situations. Usually, this form of legitimacy is almost impossible to achieve, since the state often has an interest in not recognizing any "special" rights, especially when it means that more time and resources must be spent to solve a "problem."

Conclusion

A Health and Well-Being paradigm model has not yet been identified or incorporated by the EPA. By the end of Johnson's internship several years ago, she had produced a "Literature Review and Report" outlining different tribes that practiced the paradigm's concepts. That paper was supposed to act as a starting point in creating the foundation for the model; however, no further action has occurred. The new paradigm seems to have made no progress beyond what previous conferences and meetings have accomplished. Ideas and thoughts on the paradigm continue to be discussed; however, a concept paper, model, or further documents have not been produced by the EPA, although a summary of activity was published in 2006 (U.S. Environmental Protection Agency 2006a). The discussions among tribal representatives have, however, led to further development of the models at the tribal level—work continues within Tribal Nations to define risk, environmental health, and regulatory models consistent with the themes of the Health and Well-Being paradigm even if not currently (or yet) represented in official EPA policy.

What can we learn from all of this activity? The Health and Well-Being paradigm calls on tribal perspectives that include tribal beliefs, culture, and lifestyles. For many Native and non-Native environmental professionals, the concepts in the Health and Well-Being paradigm illustrate the need to modify the risk-assessment procedure to include how Native people, especially those with nonsuburban lifestyles, are exposed to pollution—it identifies the many effects that pollution can have, and has had, on Native communities and their lifeways. Shifting to this paradigm would also require the EPA to use another form of knowledge and, most importantly, appropriately use the knowledge in the way it is intended by the tribes and the Native Americans who define it. In the politics of recognition, this is a difficult process; the liberal state is much more comfortable with the homogenized models of conventional risk assessment. Also, when it does recognize the needs of others, the state usually reaffirms its own power by controlling the categories and processes of participation. Those homogenized models remain an obstacle to achieving the opportunities made possible by the work of the TSC.

Placed in more general terms, this chapter demonstrates that EJ communities, their activities, and their critiques have started to impact the way environmental risk assessments are understood, and to a certain extent, conducted. As our experiences showed, however, a critique can easily be co-opted by the state and not serve the intent of those who put it forward. In a deeper way, the reasons for this failure of integration of the Health and Well-Being model were there all along: most of the tribal participants in the TSC wanted a different form of participation to go along with a different form of science. Adopting the Health and Well-Being approach would require a paradigm shift that would also undermine the EPA's administrative authority to control the process of knowledge production itself. While a group of scientists and managers at the EPA tried in good faith to create new, more environmentally just measures, there were limits to their efforts that were frustrating to observe. As Jasanoff (1995, 290) has noted, informal regulatory bodies like the Tribal Science Council have the most hope in "negotiating differences over 'facts' and values," at least compared to more adversarial, courtlike settings. It is clear that the Tribal Science Council allowed for "multilateral exchange, with opportunities for give-and-take between experts, the agency, and other interested participants" (p. 290), but this was not enough to affect the kind of change called for by the tribal participants on the TSC.

A true paradigm shift in something like risk assessment requires a redistribution of power, which is why defining and implementing the Health and Well-Being paradigm has become so difficult. The paradigm, a tribal approach, ideally gives decision-making power to Natives. The EPA has the power to approve programs proposed by tribes, and the EPA and its bureaucrats use their own ideals to grant Natives recognition and legitimacy. In such situations, Natives are forced, as Paul Nadasdy (1999, 1) points out, to "express themselves in ways that conform to the institutions and practices of state management rather than to their own beliefs, values, and practices." Anything else, as Johnson's experiences illustrate, is placed outside of the system and deemed irrelevant.

Thus, the blurring of lines between science, policy, and activism has led us back to more typical discussions of recognition and the functioning of liberal democracies. The violations of difference embedded in traditional risk assessment, assumptions that homogenize protections to the citizenry, are in essence the same violations of difference that eventually killed the Health and Well-Being model within the EPA—they are each problems of "recognition." While the issues of homogenization started out as what can more typically be thought of as "science" and ended in something that is more recognizable as policy, the challenges are strikingly similar. The TSC is an attempt to redesign scientific risk assessment with more meaningful attention to the affected populations and their cultural contexts. As Tribal Nations continue to do this work at the local level—be it through the TSC or otherwise—we must hope that scientists and policymakers within regulatory agencies recognize the value of this work. This recognition can lead them to write policy and practice science that are more cognizant of the ways such science and policy lead to a homogenization that affects the health and well-being of us all.

8

Uneven Transformations and Environmental Justice: Regulatory Science, Street Science, and Pesticide Regulation in California

Raoul S. Liévanos, Jonathan K. London, and Julie Sze

Introduction

The San Joaquin Valley of California is world renowned for its industrial agricultural production (Pulido 1996). Three counties in the southern portion of the Valley—Fresno, Kern, and Tulare—are consistently the top three in agricultural production and export in the United States. This prominent status of the Valley and of these three counties is facilitated by a mild Mediterranean climate; Romanesque irrigation systems and surface water delivery schemes; the industrial applications of pesticides; and the exploitation of an inexpensive, often sociopolitically isolated, immigrant farm labor population (Cole and Foster 2001; Harrison 2006, 2008; London, Sze and Liévanos 2008; Pulido 1996; Sherman et al. 1997; Villarejo 2000; Walker 2006).[1]

Understanding the environmental and human health effects of pesticide exposure in this context is increasingly a matter of concern among various regulators and advocates of pesticide reform and environmental justice. However, significant scientific disagreements exist between these parties over how to best monitor and assess the health impacts of such high pesticide use and over regulatory measures to ensure both sustainable agricultural production and public health (Eaton et al. 2008; Kishi 2005; Lee et al. 2002). One area of great contention is how best to document and regulate the adverse health effects experienced by those exposed to "pesticide drift," or "the offsite airborne movement of pesticides away from their target location" (Harrison 2006, 507).[2]

This chapter focuses on how the institutionalization of environmental justice policy and the parallel development of environmental justice movement organizing has transformed "regulatory science" (Jasanoff 1990) and "street science" (Corburn 2005) in the areas of pesticide monitoring and regulation in California.[3] In taking this approach, we add to previous literature in the following ways. First, we examine important innovations being made through regulatory and social movement efforts, whose ability to advance the goals of environmental justice are just beginning to be understood. Second, we draw on the developing New Political Sociology of Science perspective (Frickel and Moore 2006, 5), which explores the ways in which taken-for-granted rules and procedures, networks, and resources "shape the power to produce knowledge and the dynamics of resistance and accommodation that follow." Third, our analysis of transformations in pesticide monitoring and regulation extends previous work on scientific transformations involving agriculture and "university-industry relations" in the fields of biotechnology (Kleinman and Vallas 2006; Glenna et al. 2007) and cooperative extension (Henke 2006).

Through a comparative analysis of two cases, we argue that environmental justice movements have contributed to *uneven* transformations in the regulatory science and street science of pesticide monitoring and regulation in California. This transformation is contingent on the scientists' and their organizations' negotiation of their taken-for-granted notions of "science" and "environmental justice." In addition, the regulatory and social movement organizations we analyze have a history of interaction that shapes how both attempt to transform pesticide monitoring and how they conceive of precautionary, regulatory approaches to the cumulative impacts of pesticides. Finally, we highlight the opportunities and constraints imposed on regulators and advocates by the larger political economic landscape by analyzing how these uneven transformations are affected by the diffusion of neoliberalism into pesticide regulation to make it increasingly characterized by weakening regulatory coercive power and resources in favor of more voluntary, market-based solutions to environmental conflict.

Overall, our analysis illustrates the uneven process by which environmental justice movements have transformed different spheres of science as practiced by social movement organizations and some of their target regulatory agencies. Future research must continue to document how this uneven process unfolds over time to more fully understand the potential political, economic, and organizational opportunities and obstacles shaping the extent to which the environmental justice movement can transform expert cultures in a way that aligns with the goals of the movement.

Conceptual Grounding and Research Strategy

One of the key strides the environmental justice movement has aimed for is the democratization of science (Brulle and Pellow 2006; Brown 2007; Corburn 2005; Cohen and Ottinger, introduction, this volume). A critical avenue for such democratization has been through "citizen-science alliances" and "boundary movement" repertoires (Brown 2007) whereby distinctions between science and nonscience, and experts and laypeople, are blurred "in order to negotiate the meaning of science and to challenge the definitions of acceptable scientific practices and products" (McCormick, Brown, and Zavestoski 2003, 547). These alliances challenge what some see as a defining bifurcation of environmental justice struggles: identity politics based in the everyday lived experiences of grassroots community members versus the "disinterested politics" and abstracted, "objective" knowledge of technical experts (Tesh and Williams 1996).

Corburn (2005) extends these notions of citizen-science alliances by showing how their participatory research methods improve overall scientific inquiry and produce "street science." He defines street science as "a practice of science, political inquiry, and action" that "originates and evolves in a community—where community is defined geographically, culturally, or socially" (p. 44). It "distances itself from mentalistic and subjectivistic views of judging, assessing, and knowing" with its use of professional and technical expertise (p. 44). This conceptualization of street science is applicable to many cases of the negotiation of "local knowledge" and "professional knowledge," but Corburn acknowledges (p. 51) that he is conflating two types of scientific practice in the category of professional knowledge: regulatory science and research science as discussed by Jasanoff (1990).

We maintain it is important to elaborate the conceptual distinctions between regulatory, research, and street science when attempting to explain how the environmental justice movement has transformed separate spheres of science (see table 8.1). As table 8.1 shows, regulatory science is a type of science used for policymaking predominantly by government and industrial institutions (Jasanoff 1990). Research science, conducted primarily, but not exclusively in universities, is largely driven by professional recognition, career advancement, and legitimacy in the eyes of a given scientific field (Jasanoff 1990; Frickel and Moore 2006). We find both contrast with street science's social justice and rights-based discourse (Brown 2007; Cole and Foster 2001; Sze 2007) and its pattern

Table 8.1
Three spheres of science typically involved in movements for environmental justice

	Regulatory science	Research science	Street science
Truth claims	"Truths" to inform and legitimate public policy	"Truths" of originality and disciplinary significance	"Truths" relevant to the lived experiences of community members; policy advocacy; significance for action
Practicing institutions	Government; industry	Universities; government; industry	Community-based groups; SMOs; universities; government
Products	Studies and data analyses, often unpublished; private goods (with exchange value); public goods (use value and for democratic participation)	Published papers; public goods; private goods	Studies and data analyses; published and unpublished; press releases; community organizing, education, empowerment; public goods
Incentives	Compliance with legal requirements; balance interest/disinterest in economic development	Professional recognition and career advancement; legitimacy within the scientific field	Achievement of social justice and human rights; legitimize the claims of SMOs and participating community members in the public sphere
Time frame	Statutory timetables; political pressure; funding and career constraints	Open-ended; funding and career constraints	Statutory timetables; political pressure; time demands of involved community members; funding and organizational constraints
Options in the face of uncertainty	Acceptance of evidence; rejection of evidence	Acceptance of evidence; rejection of evidence; waiting for more data	Acceptance of evidence; rejection of evidence; waiting for more data; precautionary or preventive; consensus over cause not necessary

Accountability	Congress/state legislatures; courts; media; the targets of the findings	Professional peers; targets of the findings	Social movement peers; participating community institutions; courts; media; government; targets of the findings
Procedures for scientific review	Audits and site visits; judicial review; peer review (formal and informal), from "stakeholders" and "the public"	Peer review (formal and informal), from universities and research funders (e.g., National Science Foundation)	Coproduction of knowledge; broadly defined "peer" review including nonscientists, SMOs, universities, regulators
Methodological and ethical standards	Absence of fraud or misrepresentation; conformity to approved protocols and agency guidelines; legal tests of sufficiency (e.g., substantial evidence, preponderance of the evidence)	Absence of fraud or misrepresentation; conformity to methods accepted by peer scientists; statistical significance	Absence of fraud or misrepresentation; alignment with the social movement principles and community expectations; legal tests of sufficiency; triangulation of statistical significance with lived experience of community members
Boundary actors, practices, norms, and objects	Blur and reify the boundaries of each sphere of science (e.g., university and regulatory scientists as boundary actors in each sphere; epidemiology and toxicology as a boundary practice; environmental health and justice as boundary norms; various monitoring equipment as boundary objects).		

Compiled from the empirical cases in this chapter and from Brown 2007; Cole and Foster 2001; Corburn 2005; Frickel and Moore 2006; Glenna et al. 2007; Gould 1994; Jasanoff 1990; Keck and Sikkink 1998; Kleinman and Vallas 2006; McCormick, Brown, and Zavestoski 2003; Pellow 2001, 2007; Star and Griesemer 1989; and Sze 2007.

of drawing on scientific practice to legitimize the claims of social movement organizations (SMOs) and community members in public debate (Keck and Sikkink 1998).

In the cases that follow we conduct a comparative analysis of the uneven transformations in regulatory science and street science in relation to the classifications in table 8.1.[4] First, we explore how the scientific practice of the California Department of Pesticide Regulation (DPR)—a widely recognized national leader in pesticide regulation (Harrison 2008)—responds to challenges of legislative mandates to engage with environmental justice principles of "fairness" and methods of cumulative impact assessment, precautionary regulation, greater public participation, and community capacity building. We analyze how all of these components were at play in DPR's "Environmental Justice Pilot Project" in the Fresno County town of Parlier, California. Second, we analyze these components in the street science practiced by a local-to-global network of pesticide reform and environmental justice advocates who combined community-based participatory human biomonitoring, or "grassroots body burden testing" (Brown 2007), with pesticide drift catchers to coproduce what they term "*la realidad*": the everyday "reality" of coping with the burden of living, working, and playing amid pesticide-laden air, soil, and water (see also Morello-Frosch et al., chapter 4, this volume). This coproduction of knowledge was conducted through both the statistical and narrative construction of the "unacceptable" levels of pesticides in the bodies and indoor air of Latina/o residents in the Tulare County town of Lindsay, California.

For simplicity, we refer to the Parlier pilot project as the "Parlier Collaborative" and the project in Lindsay as the "Lindsay Collaborative." Figure 8.1 shows the geographic location of each place within the San Joaquin Valley and in relation to neighboring places. The figure also shows how the block group percentage of persons eighteen years and over with a 1999 income below the poverty level and who are Latina/o in Parlier and Lindsay is higher than their respective county averages.[5] Both collaboratives conducted all or portions of their pesticide monitoring in 2006 and were designed to engage disproportionately poor, Latina/o agricultural worker communities and their advocates. Each collaborative was also designed with specific environmental justice and precautionary principle–related policy outcomes in mind. Yet, each differed greatly in their origins, monitoring methodology, orientation to science and environmental justice, definition and role of community, and outcomes as summarized in table 8.2.

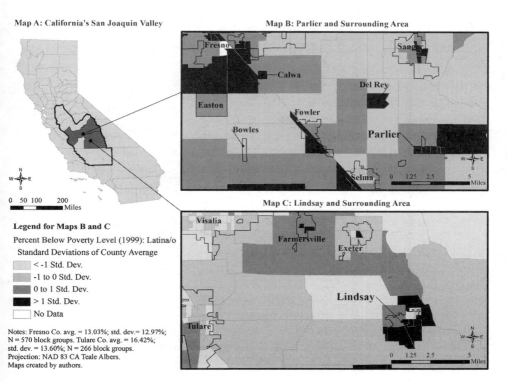

Map A: California's San Joaquin Valley

Map B: Parlier and Surrounding Area

Map C: Lindsay and Surrounding Area

0 50 100   200
Miles

Legend for Maps B and C

Percent Below Poverty Level (1999): Latina/o
Standard Deviations of County Average

< -1 Std. Dev.
-1 to 0 Std. Dev.
0 to 1 Std. Dev.
> 1 Std. Dev.
No Data

Notes: Fresno Co. avg. = 13.03%; std. dev.= 12.97%;
N = 570 block groups. Tulare Co. avg. = 16.42%;
std. dev. = 13.60%; N = 266 block groups.
Projection: NAD 83 CA Teale Albers.
Maps created by authors.

Figure 8.1
Percent 1999 income below poverty level and Latina/o in the study areas

To guide our inquiry, we asked three primary research questions: (1) How do different organizations attempt to implement environmental justice principles and practices as they practice their science? (2) How have institutional opportunities and constraints shaped such practice? (3) How have those practices evolved due to interaction with the environmental justice movement? We conducted eighteen in-depth, semistructured interviews (from 45 minutes to 2.5 hours in length) with the lead scientists and researchers in the Parlier and Lindsay collaboratives, top DPR management, and participating community members and pesticide reform and environmental justice advocates. We analyzed organizational documents to assess how the organizations involved in the collaboratives were addressing environmental justice and their scientific practice over time. Furthermore, we observed twelve DPR stakeholder meetings and three demonstrations and press conferences held by environmental justice advocates pertaining to air quality issues and pesticide reform,

Table 8.2
Summary of the two collaboratives

Collaborative	Origins	Monitoring methodology	Orientation to science; environmental justice	Definition/role of community	Outcomes
Parlier	California legislation and administration of the California Environmental Protection Agency	Fixed monitoring stations for 40 pesticides, 3 times per week over 12 months	Regulatory science; environmental justice as legal mandate and orders from agency leaders	"Stakeholders" invited as part of "Local Advisory Group"	Unprecedented monitoring and participation protocols for DPR; detailed but delayed monitoring report; further study of pesticides of concern; continued focus on voluntary, reduced-risk pesticide management
Lindsay	Local-to-global network of pesticide reform and environmental justice advocates	Human biomonitoring of community members and 6 drift catchers around their residences for 1 target pesticide during peak spraying periods over 3 years	Street science; Citizen-scientist alliance to stop exposure to pesticide drift on a low-income, Latina/o community	Those directly affected on the "fenceline" empowered to speak and prove for themselves with pesticide reform and environmental justice advocates	Successful political pressure for new county buffer zones and pesticide spray regulations in Tulare County and three neighboring San Joaquin Valley counties

and two regional air quality public hearings. We now turn to the insights derived from our conceptual groundings and research strategy.

Transforming Regulatory Science: The Parlier Collaborative

The Parlier Collaborative originated from a series of statutes to create a comprehensive state-level regulatory approach to environmental justice in California. California's comprehensive environmental justice legislation began in 1999 with the legal definition of environmental justice in state code as: "the fair treatment of people of all races, cultures and income with respect to development, adoption and implementation of environmental laws, regulations and policies."[6] By 2004, environmental justice policy was coordinated through the California Environmental Protection Agency (Cal/EPA) Environmental Justice Action Plan. The Action Plan applied to all of Cal/EPA's agencies, including DPR. DPR was charged with implementing a pilot pesticide monitoring project to assist Cal/EPA in developing cumulative impact analyses of pesticides, precautionary approaches to address such impacts, improved tools for public participation, and community capacity building in the process. It also stipulated that there be an analysis of potential health risks posed to children. These elements were to comply with the governor's "Environmental Action Plan" for California's compromised land, air, and water resources (Cal/EPA 2004).[7]

The push to integrate cumulative impact and precaution into the sphere of regulatory science in Cal/EPA's agencies is largely due to the organizing efforts of the environmental justice movement at the state and national level. As advanced by the movement participants, the notion of cumulative impacts seeks to recognize the multiplicity of environmental threats on typically low-income communities and communities of color over time. Movements for environmental justice have also been at the forefront of pushing for precautionary measures to combat such cumulative impacts (Brulle and Pellow 2006; Corburn 2005). This effort is based on the principle that "where there are threats of serious or irreversible damage, the lack of full scientific certainty shall not be used as a reason for postponing cost-effective measures to prevent environmental degradation" (Whiteside 2006, viii).

Aligning with Environmental Justice

The ability of DPR's personnel to accommodate the concerns of environmental justice advocacy and legislation is shaped by DPR's self-described "science-based" character with the mission "to protect human health and the environment by regulating pesticide sales and use, and by fostering reduced risk-pest management" (CDPR 2007b). Speaking to that issue, the administrators and scientists we interacted with told us how difficult it is to negotiate some of their personally limited understanding of environmental justice with the legal definition and DPR's mission. For example, DPR Director Mary-Ann Warmerdam (2007a)[8] described the challenge in these terms: "I think one of the *challenges* that we have as a public agency is to try and understand and appreciate how the advocates of [the legal] definition [of environmental justice] feel about the work that we do. . . . And I think that's where it starts becoming very difficult for us to *define* environmental justice [in a way] that makes sense to them and to us as well" (emphasis in original). Likewise, one twenty-six-year veteran toxicologist, who worked on the Parlier Collaborative, summed up the unfamiliarity of some agency scientists with environmental justice principles and how engagement with the movement, particularly its community or place-based focus, would change their work: "I had not been aware of this particular issue prior to this state law being passed and then subsequently with [various organizational plans] and things like that. We don't often look at specific communities and see what their exposure is. So, that's a change in perspective for us" (DPR Scientist 1, interview July 24, 2007, Sacramento, CA).

In contrast, administrators and scientists more familiar with the history of the environmental justice movement saw DPR's recent policy change as something that fit in with what the organization had already been doing in its work, particularly its monitoring. One scientist with sixteen years' employment in DPR commented:

I remember hearing about it a long time ago when issues like Love Canal, or a lot of industrial things were going on in . . . lower income and poverty areas [where] *people didn't have the power* to control what was going on. . . . I think also a lot of our studies are directed towards areas of high use of pesticides, which in a lot of ways ends up being some of these communities that are environmental justice communities 'cause they're out there in the middle of the Ag lands and stuff. . . . So, I think that [environmental justice has] kind of *been there*, but without a name or anything to it. (DPR Scientist 2, interview July 24, 2007, Sacramento, CA; emphasis in original)

A similar sentiment was conveyed to us from elsewhere in the organization when referring to how DPR's programmatic areas of enforcement, environmental monitoring, pest management, licensing, registration, and especially worker health and safety can all fit under the umbrella of environmental justice—if framed correctly. As one DPR official noted, "an easier way to understand it when I explain it to our staff, our management team, is that part of the mandate of the law is that we analyze our programs and policies to see what barriers there might be to the achievement of environmental justice. People have a little easier time understanding EJ when they see it in that context" (DPR Official 1, interview, Sacramento, CA, May 15, 2007). What emerges from these voices is that with the passing of environmental justice legislation and policies over the past decade, DPR and its personnel are now required to negotiate and implement a definition of environmental justice that complies with the law and is compatible with their personal knowledge and their organizational contexts.

Cal/EPA's multistakeholder deliberations resulted in Cal/EPA's working definitions of cumulative impact and precaution in 2005 that would soon influence DPR's scientific practices.[9] Per agency policy, these two concepts were to be integrated into each of Cal/EPA's pilot projects when conducting cumulative impact analysis and drafting precautionary regulatory approaches to address such impacts. When reflecting on how each definition would influence the regulatory science conducted by DPR, a lead scientist in the Parlier Collaborative commented to us: "Well, the precautionary approach is more of a policy issue, so I'll stay away from that one. In terms of the cumulative impacts, it does change probably the scope of what we're looking at" (DPR Scientist 1, interview July 24, 2007, Sacramento, CA).

This comment was representative of two themes that arose among DPR scientists and administrators affiliated with the Parlier Collaborative. First, they expressed a form of boundary work when they spoke to perceptions of what is in the realm of their scientific practice and what is not (Gieryn 1983): the scientists strive to focus strictly on "the science" of cumulative impact assessment, while the administrators develop the policy or "nonscience" of precautionary risk management. Yet even this "nonscience" is informed by the scientists' work. Second, this comment represents how boundaries shifted within DPR about what type of monitoring activity they should and legally can conduct. DPR only has legal authority to monitor and regulate pesticides, and has customarily done so by assessing pesticide toxicology in only one environmental media at a time (i.e., in soil, water, or air). As a result of legislation and associated agency regulations developed based on pressure from the environmental justice advocates and their allies in the legislature, DPR and other state agencies are now grappling with how to look at a whole range of exposures as a result of the increasing focus on cumulative impact and precaution.

Innovations in the Parlier Collaborative

State environmental justice legislation pushed DPR to adjust their channels of accountability to include greater public review and scrutiny of their procedures and methodological considerations for selecting a monitoring site in the Parlier Collaborative. This adjustment is limited when compared to potential avenues of scientific transformation (see table 8.1), but at least it resulted in increasing their formal recognition of the social structural factors (race, class, etc.) associated with the disproportionate exposure to environmental threats. It also resulted in the simultaneous effort to "scale down" DPR's statewide focus to a community-based intensive methodology to monitor a multiplicity of pesticides in place for an entire year. These transformations pushed DPR to refine how it integrates its existing work to support voluntary integrated pest management that attempts to protect public health and the environment in accordance with Cal/EPA's working definition of precautionary approach (Matteson, Wihoit, and Robertson 2007).

Cal/EPA's working definition of cumulative impact gave DPR scientists the opportunity to focus on selecting monitoring sites based on a matrix of criteria. DPR assessed potential monitoring sites based on three categories: (1) community "environmental justice factors," which include percent population less than eighteen years old, percent nonwhite population, median family income level, and the number of pesticide-drift illnesses; (2) the "availability of cumulative impact data," such as DPR's pesticide well-water monitoring and monitoring stations for U.S. EPA–defined criteria air pollutants;[10] and (3) regional pesticide use (within five miles of a community) and local use (within one mile of a community) (CDPR 2005). These criteria were part of DPR's effort to maintain "objective" standards as compared to an earlier community monitoring project.[11] As one DPR scientist told us, "We did try to select Parlier based as much on objective criteria as we could. We specifically avoided just picking at random, or based on somebody's suggestion. We didn't want a subjective selection, so we set the scoring system and based on that system, Parlier came out on top" (DPR Scientist 1, interview July 24, 2007, Sacramento, CA).

The use of environmental justice indices to identify communities disproportionately exposed to environmental threats in urban areas has diffused to regulatory agencies across the United States (Harner et al.

2002). To the best of our knowledge, DPR's selection criteria were the first used by a regulatory organization in the country in the context of an environmental justice–motivated agricultural pesticide monitoring project. Despite the pressure by pesticide reform and environmental justice advocates to select another community with more of an organized environmental justice movement presence, Parlier was selected because it ranked highest of the eighty-three communities evaluated by DPR. The lack of a community base notwithstanding, Parlier's selection was a welcome recognition of a town exemplifying the San Joaquin Valley's pattern of "poverty amidst prosperity" (Martin and Taylor 1998; see also figure 8.1) and farmworkers' "suffering in silence" (Sherman et al. 1997; Villarejo 2000).

The second and perhaps more significant innovation in the Parlier Collaborative was the integration of formal public engagement in the monitoring process. DPR still maintained ultimate authority over each aspect of the project and drew on its established practice of peer review by regulatory and research scientists through the use of technical advisory committee. However, DPR's chartering of a local advisory group within the Parlier Collaborative was the first attempt by DPR to integrate public scrutiny and local knowledge into its scientific protocol. Such practice has been the typical model used by Cal/EPA agencies in their various pilot projects and has been met with mixed results (London, Sze, and Liévanos 2008). Yet when compared to others used in the pilots, this local advisory group engaged regional and statewide environmental justice advocates and had its local knowledge explicitly valued and used to scrutinize and inform participating scientists in the locations where DPR and their partnering organizations had placed their air monitors in the community.[12] One scientist working closely with the collaborative added: "They have impacted how we've conducted the monitoring, where we've monitored and things like that. . . . I know a lot of times we don't think of those things because we're kind of immersed in it" (DPR Scientist 2, interview July 24, 2007, Sacramento, CA).

This break with standard practice—or "rupture," to draw on the language of this book—led to a new relationship with public stakeholders. That new relationship presented DPR scientists with a situation in which they had to confront the disciplinary stigma about the collaborative with the higher level of personal fulfillment it gave them. This sentiment was expressed well by one DPR scientist, who understood that the collaborative was seen as "not a sexy type of research project; it's more applied research, rather than cutting edge stuff." But, in the next breath, this individual went on to note that the extended study in the field was "more fulfilling" because scientists were able to work with some members of the communities, which contrasts with their typical model of "just grabbing your people, going out some place, measuring, and then coming back to take your little data and put it into your computer for modeling" (DPR Scientist 2, interview July 24, 2007, Sacramento, CA).

Despite being "not sexy," DPR's community engagement and associated practices of scientific production gave rise to the most significant transformation in the regulatory science of DPR: it opened up the "black box" that was the organization's regulatory science. Pesticide reform and environmental justice advocates noticed a change in DPR by how open the administration and the scientific staff had become to public comment during and following the Parlier Collaborative. For example, one advocate commented: "DPR now, overall, appears to be a lot more open and more interested in taking comments and more interested in actually going out to seek comments from community members" (EJ Advocate 1, interview June 4, 2007, San Francisco). However, it should be noted that while DPR did open itself up to public scrutiny and community engagement, it also continued to retain ultimate authority over the collaborative, its machinations, and the development of subsequent policy recommendations based on the findings (London, Sze, and Liévanos 2008).

After a year of monitoring, the Parlier Collaborative found twenty-three pesticides, or their breakdown products, present in Parlier's ambient air, including two pesticides that exceeded the nonregulatory criteria of "health screening levels": diazinon and acrolein. Acrolein is a widespread breakdown product and environmental pollutant (Szadkowski and Myers 2008) and was attributed by DPR to nonpesticidal sources, such as vehicle emissions (CDPR 2007a). Diazinon, a toxic organophosphate pesticide, has been placed on DPR's high-priority list, and is currently undergoing risk assessment by the organization. Chlorpyrifos and one of its breakdown products were found in Parlier's air. This organophosphate pesticide has been linked to potential neurodevelopment problems and is metabolized in adults within hours (Eaton et al. 2008). Chlorpyrifos was introduced into the global market in 1965; it has been and continues to be used in agricultural production, though it was banned from residential use in the United States in 2001 (Eaton et al. 2008). DPR scientists maintain that the screening level for chlorpyrifos was not exceeded in Parlier, but DPR's protocol for risk assessment did not call for using the recent Food Quality Protection Act (FQPA) of 1996 safety factor for evaluating the risk of chlorpyrifos to infants and children in developing the screening level (CDPR 2007a; Lee et al. 2002). If they had done so, 6 of the 468 samples they collected of chlorpyrifos would have exceeded the acute (one-day), noncancer health screening level (CDPR 2007a).

Despite this evidence, DPR postponed the final report for some clear and not-so-clear reasons. This is an element of the "science game" pesticide reform and environmental justice advocates are weary of in their interactions with DPR and in their general skepticism of DPR's regulatory science. An October 2007 letter from DPR Director Warmerdam to the Parlier local advisory group explains *some* reasons behind the delay in issuing the final report (finally released in late 2009). In doing so, it represents the favorable "presentation of self" (Goffman 1959) by DPR leadership in potentially contentious, public debates about its regulatory science:

We find that we must delay [the] report, perhaps as late as the first part of 2009. DPR's air monitoring staff (which will write the Parlier project report) has been assigned the huge task of implementing new regulations designed to reduce smog-producing emissions from fumigant pesticides. . . . Delaying the report will not affect what has been achieved in the Parlier project. The project goal is to evaluate ambient air exposure to pesticides to better understand and identify opportunities to reduce environmental health risk, particularly to children. The data do not show significant health risks that warrant immediate action. As promised, however, we are taking a closer look at findings that were above or close to our health-protective screening levels. We are not waiting for the final report to do that. (Warmerdam 2007b)

What is not stated as clearly in this quote is how, as a regulatory science–based organization, DPR's scientific practice is heavily influenced by legal requirements and, as will become evident later, the imperative to balance its interestedness in the political economy of industrial agricultural production. Additionally, its regulatory science is susceptible to its interactions with the social movement organizations that target them. For example, in revisiting its decision not to apply the FQPA safety factor in conducting its health screening for chlorpyrifos, the final report released at the end of 2009 found chlorpyrifos (along with diazinon) as contributing a disproportionate amount of relative health risk, and therefore warranting a higher priority for future risk assessment than was previously planned (Wofford, Segawa, and Schreider 2009). Besides being hampered by the funding limitations on the Parlier Collaborative (discussed later in the chapter), the postponement of the report stemmed from limited agency resources coupled with a lawsuit filed against DPR by pesticide reform and environmental justice advocates over stricter regulations of fumigant pesticides. These type of pesticides volatilize into the air (as volatile organic compounds, or "VOCs"), contributing in that manner to the formation of ground-level ozone, or "smog." This concern over VOCs and smog has become a considerable focus and hotbed for environmental justice organizing in the San Joaquin Valley.[13]

Thus, the pursuit of stricter regulatory attention to VOCs by advocates contributed, in part, to DPR releasing its final report on pesticide monitoring in Parlier later than initially promised. However, advocates' efforts in Lindsay showed DPR how to conduct a more community-empowering and scientifically credible analysis of the disproportionate risk of chlorpyrifos application on vulnerable populations, such as low-income, mostly Latina/o farmworkers (Wofford, Segawa, and Schreider 2009).

Transforming Street Science: The Lindsay Collaborative

In addition to using "bucket brigades" or relatively inexpensive air-monitoring devices to capture contaminants in the environment of low-income communities and communities of color, some community groups are starting to participate in grassroots body-burden testing. This new repertoire derives from "critical epidemiology" and citizen-science alliances for environmental justice (Brown 2007). As Brown notes, critical epidemiology values the democratic production and distribution of knowledge, and "looks at the historical context of epidemiologic investigations, takes a social structural and health inequalities approach, and applies the discipline's epistemology for social justice by working with laypeople or by seeing themselves as tied to social movements or by doing both" (p. 263). This increasing interest in environmental justice movements on chemicals inside and outside the body signals that there is a transformation underway in the practice of street science (also see Morello-Frosch et al., chapter 4, this volume). As we outline in more detail below, the associated methodological and ethical standards around community-based participatory research are spurring the innovative use of boundary practices (grassroots body-burden testing and critical epidemiology), norms (community empowerment), and objects (the "drift catcher") (see table 8.1 for more).

The Lindsay Collaborative was a citizen-science alliance that sought to document the body burden of chlorpyrifos and its presence in the indoor air of Latino residents in the Tulare County town of Lindsay. The focus on the presence of this pesticide, and its associated breakdown products, in the bodies and indoor air of community members suggests that aerial agricultural application could be responsible for the documented levels of chlorpyrifos. Developed between 2003 and 2006, the Lindsay Collaborative joined members from Commonweal, the Pesticide Action Network of North America (PANNA), Californians for Pesticide Reform (CPR), and the participating Lindsay community group, *El Quinto Sol de América*.[14] These four key groups came together as pesticide reform advocates were raising awareness about pesticide-drift exposure in California. One of their first major victories came in 2004 when sympathetic legislators signed legislation the collaborative advocated for that increased coordinated regulatory responses to pesticide drift in nonoccupational settings and that would hold parties responsible for injury or illness liable for uncompensated medical care (London, Sze, and Liévanos 2008). Their project planning also occurred while the organizations were involved in passing legislation to create the California Environmental Contaminant Biomonitoring Program in 2006. The legislation was the first of its kind in the nation and helped create national awareness about the importance of making corporations and regulators accountable for harmful patterns of "chemical trespass" on low-income people of color who live on the "fenceline" near environmental pollution (Brown 2007; Lerner 2005; Schafer et al. 2004).

The collaborative represents the kind of participatory exercise of environmental justice that challenges the "dominant epidemiological paradigm" (Brown 2007), which predominantly focuses on national populations of the exposed while not involving those who are monitored in the data collection, analysis, and/or presentation of the results. Those in the Lindsay Collaborative believe they must "work with the most directly affected people and bring them up into leadership positions where they can speak and advocate for themselves" (EJ Advocate 2, interview June 4, 2007, Oakland, CA). The collaborative built on their shared principles, recent legislative victories, ongoing advocacy around pesticide drift in Tulare County, and increasing legitimacy given to biomonitoring in exposure assessment to document the "unacceptable levels" of body burden in a low-income community of color (Schafer et al. 2004). This unacceptable level documented through the collaborative's street science was used to push the Tulare County Agricultural Commissioner, the local enforcement arm for DPR, to adopt a precedent-setting precautionary approach in land use policy in pesticide-drift prone areas. The approach calls for an aerial application prohibition of restricted-use pesticides in a one-quarter-mile buffer zone around schools within 24 hours to being in session, occupied farm labor camps, and residential areas in the county effective January 1, 2008. While this approach has not been successfully legislated at the state level, similar forms of this policy have diffused to other San Joaquin Valley Counties (Madera in 2009 and Stanislaus and Kern in 2010).[15]

Aligning with Environmental Justice

PANNA, Commonweal, and CPR have historically been organizations led by a predominantly white, affluent staff that have used the discourses of the antitoxics and "environmental reform" movements (Brulle 2000) rather than movements for environmental justice. Their work has focused on coordinating statewide and national efforts to ban some of the most harmful pesticides used in industrial agriculture. These three organizations also push for alternatives to the conventional risk assessment conducted by regulators and industry groups because they feel it lacks a proper attention to cumulative impacts and the precautionary principle. However, the character of this advocacy has drastically changed in the past decade as they have made efforts to partner with low-income communities and communities of color and change their personnel to reflect the demographics of the environmental justice movement.[16]

PANNA, Commonweal, and CPR have brought low-income individuals and people of color into powerful roles in their organizations and have devoted significant economic and human resources to participate in environmental justice advocacy as a result of their engagement with environmental justice and other community-based groups. For example, CPR has hired the founder of *El Comité Para El Bienestar de Earlimart*, another Tulare County community group, who is one of the contemporary community-based leaders in organizing against pesticide drift in the San Joaquin Valley, to be the lead representative for CPR in the region. CPR has also been hiring other local organizers and leaders to help build the capacity of the communities most affected by pesticides. Furthermore, taking this sort of action has aligned a significant amount of CPR's organizational mission with those of the environmental justice movement. CPR initiated the "Safe Air for Everyone (SAFE)" campaign to integrate their platform into land-use policy that would change fenceline communities into buffer zone–protected places in Tulare County. In doing so, they sought to innovatively articulate *la realidad* of pesticide contamination, with the grassroots organizing of *El Quinto Sol de América,* as discussed below.

Innovations in the Lindsay Collaborative

A key transformation in the street science practiced in the Lindsay Collaborative was the change in its methodological and ethical standards around community-based participatory research, which necessitated the innovative combination of boundary practices, norms, and objects in the project. The key boundary practices of grassroots body-burden testing and critical epidemiology implemented in the collaborative drew on the combination of community residents' storytelling in public venues about their experiences of pesticide drift with PANNA and Commonweal's epidemiological approach to the use of human biomonitoring and pesticide air monitoring. The air monitoring was facilitated by a key boundary object: the drift catcher. Developed by PANNA scientists with technical assistance from regulatory scientists from DPR, its design follows protocols developed by the National Institute for Occupational Safety and Health and the California Air Resources Board. The drift catcher is essentially a collection bag and pump that sucks pesticide-contaminated air through tubes packed with an absorbent resin to catch the pesticides for subsequent analysis (CPR 2007b). The key boundary norm weaved throughout the project was community empowerment to fight the injustice of unacceptable levels of chemical body burden and air pollution.

The Lindsay Collaborative released its results on May 16, 2007, on a house lawn across the street from a local school as part of the CPR SAFE Campaign. At the beginning of the press conference the *El Quinto Sol* organizer introduced the report, titled *Airborne Poisons: Pesticides in Our Air and in Our Bodies* (CPR 2007b). With press cameras focused in, she told everyone present that the collaborative has shown how Lindsay's air is contaminated with chlorpyrifos due to its heavy aerial application in the orchards around their schools, communities, and families. She added, "[We did] this project because we see the necessities. *We tell and we prove* how . . . these . . . pesticides affect our health. . . . We have the results and we have the scientists to explain a little more about that. But the main point is: today, we want to tell everybody who breathes the air, we have the responsibility to clean our air" (emphasis in original). A PANNA scientist followed and gave her presentation of the study and its results:

In 2005, nearly 2 million pounds of chlorpyrifos were applied in agricultural fields in California. For three years we did drift catching, or air monitoring, in homes in the Lindsay area. . . . In the three week period, one-third of the houses [had] levels above what the [U.S.] EPA considers safe for exposure for a one year old child. Then, for one week of that three-week testing period in 2006, 12 individuals collected their urine to say, "If this pesticide is in the air we breathe, is it in our bodies too? We think it is, and we want to demonstrate that." And, in fact, the resounding result is yes. And here we have a brief illustration of the data from the urine testing, from the biomonitoring.

The PANNA scientist then pointed to the graph and added that the U.S. EPA has determined an "acceptable" average level of chlorpyrifos that is safe for pregnant and nursing women and other adults. The collaborative's findings show that all but one of the twelve people sampled in the study (eight women and four men) had body burdens of chlorpyrifos above that average. Individual testimonies from community members began after the PANNA scientist reviewed the results. Each testimony contained a narrative about how Lindsay residents or their family members have experienced pesticide drift at their schools, near their homes, or where they work. One woman, accompanied by a Spanish-to-English translator, discussed an incident at a Tulare County school where pesticides were reportedly sprayed outside while children were playing, causing some children to collapse. A male participant followed, echoing some of these concerns: "It's not fair that our air is this bad, and pesticides are in our bodies. . . . We don't choose to breathe, but we have to."

Despite the scientific rigor displayed in this study, community members participating in the collaborative claim that local regulatory authorities frame them as "common people, not scientists," in order to question the study's scientific validity (EJ Advocate 2, interview June 4, 2007, Oakland, CA). This criticism echoed critical statements a year earlier in a local newspaper. In this article, the Tulare County Agricultural Commissioner was interviewed about the scientific standards of that study and was quoted as saying: "We will not know whether or not this represents a threat until the data is reviewed by a group of scientific experts in the field of air monitoring" (Burgin 2006). In a significant turnaround, this commissioner ultimately supported the advocates' proposal for a spray buffer around homes and schools after the study was reviewed and validated by experts in air monitoring.

The Lindsay Collaborative used street science to legitimize the claims of its participating organizations about *la realidad* of the chemical trespass in the homes and bodies of Lindsay's residents, resulting from the agricultural application of chlorpyrifos in nearby orchards. One of the study's scientists reflected on this successful collaboration: "The community was just beautifully engaged at the end. . . . Not everybody was outspoken, and that's fine. Being flexible about protocol and project design and questions asked, I think, was crucial for making the project as successful as it has been" (EJ Advocate 2, interview June 4, 2007, Oakland, CA). Lindsay's "beautiful engagement" in the study was aided by the shared orientation in the collaborative about community empowerment and the injustice of chemical trespass. It was also facilitated by the community advocates' efforts to speak and prove for themselves how their everyday living conditions are fundamentally different from those of people not living with current and potential exposure to pesticide drift.

The Lindsay Collaborative suggests there is a transformation underway in the methodological and ethical standards of street science for environmental justice. Opportunities for this transformation come from the innovative use of boundary practices (grassroots body-burden testing and critical epidemiology), norms (community empowerment to speak and prove for themselves), and objects (the drift catcher). Looking across the Parlier and Lindsay collaboratives, we see the cases illustrating a key theme of this book: how movements for environmental justice have powerfully—if unevenly—transformed scientific practice. Furthermore, the institutional and network conditions within each collaborative provided both opportunities and constraints that contributed to this uneven process. We now turn to how broader political economic conditions shaped the two collaboratives to better understand this process of uneven transformation.

Uneven Transformations across Neoliberalizing Spaces

The ability to assess cumulative impacts of pesticides, develop precautionary approaches to such impacts, enhance public participation, and build community capacity in the two collaboratives was also shaped by a broader factor stemming from ongoing dynamics of global political economic restructuring. Such restructuring is often associated with neoliberalism—an ideological and political economic project to favor voluntary and market-based economic, environmental, and social policy (Harvey 2006; Pellow 2007). Harrison (2008) claims that over time, the pesticide regulatory structure of California and the United States as a whole has experienced a neoliberal turn with its marginalization of coercive state enforcement and increasing focus on voluntary and market-based measures to register and even promote "reduced-risk" or "less toxic" pesticides into agricultural production systems (see also Pellow 2007).

This market-based orientation marginalizes the commitment to eliminate the use of toxic pest management practices and therefore to restrict capital's pattern of "capitalizing on environmental injustice" (Faber 2008). Indeed, two-thirds of DPR's pesticide regulatory program is provided by a "Mill Fee" used to tax pesticide sales and use in California, which means the agency depends on the continuation of large-scale pesticide use for its own survival. Advocates have expressed concern about this conflict of interest on the part of a regulatory organization charged with assessing risk and managing it, thereby being "captured by industry." DPR has also been viewed with some distrust by some of its "sister" agencies in Cal/EPA for being "the poison people," because they are the ones helping to facilitate the introduction of pesticides into the environment, while their colleagues are aiming to counteract potential harm caused to the land, air, water, and human bodies carrying these pesticides.

Pesticide reform and environmental justice advocates view this neoliberal market for pesticide regulation as a fundamental barrier to DPR's ability to accomplish any form of environmental justice because, as they observe, it takes the notion of eliminating the use of toxic pesticides off the table. This was the case in DPR's multistakeholder discussions about implementing environmental justice principles and practices in the organization while it was carrying out the Parlier Collaborative (London, Sze, and Liévanos 2008). Furthermore, the devolved structure of pesticide regulation in California has made county agricultural commissioners—entities who are appointees of local political and economic elites (County Boards of Supervisors)—responsible for local enforcement over pesticide drift (Harrison 2008). When it comes to the use of science, this regulatory structure is critiqued for being more interested in using regulatory science and research science as a research-and-development division for industrial agriculture than in integrating the precautionary principle into its assessment and *reduction* of environmental risk (Harrison 2008).

This system of market-mediated relationships compromised DPR's ability to develop the Parlier Collaborative and its broader efforts to open up its regulatory black box to the public in the name of environmental justice. For example, while DPR scientists had originally hoped to develop a statewide network of community monitoring sites, a series of massive budget cuts beginning in the fiscal year 2001–2002 (Harrison 2008), coupled with the high costs of the extensive monitoring protocol, limited the selection to just one site. Meanwhile other Cal/EPA agencies were able to develop similar collaboratives in more than one place (London, Sze, and Liévanos 2008). DPR scientists and staff had to use their *entire* two-fiscal-year monitoring budget of $200,000 for one calendar year of monitoring in the Parlier Collaborative.

This lack of funds signifies both the prioritization conferred on the Parlier Collaborative by DPR—an example of regulatory science—and the relative degree of budget deprivation neoliberalism imposes on an entity like DPR. Reflecting on this situation, a DPR scientist coping with "decisions from above" and with the lack of funding stated that environmental justice advocates watching the Parlier Collaborative would be justified in being left dissatisfied and that in a perfect world DPR would have developed a completely different project. This scientist imagined the reaction of pesticide reform and environmental justice advocates to the agency's work:

I'm sure they will say that we were making a good but incomplete effort. Due to a limit in resources, we were unable to do everything that they wanted us to do, but I'm sure they would make that comment and complain about it and rightfully so . . . I mean, if they had their way we would be monitoring five to ten locations rather than three [within a community], we'd be monitoring seven days a week, rather than three, we'd be monitoring for one hundred pesticides rather than forty. . . . Well in a perfect world, we'd be taking samples [of food] out of the grocery stores . . . [in addition to air, ground- and surface water]. . . . We would also be doing biomonitoring [of] the people in Parlier. . . . Again, that's a whole level of complexity that we weren't prepared to undertake. (DPR Scientist 1, interview July 24, 2007, Sacramento, CA)

The Lindsay Collaborative was able to fill the void left by DPR's lack of resources, organizational capacity, and political will. The ability of the Lindsay Collaborative to conduct its own monitoring and provide public policy recommendations for land-use and pesticide-use reform was greatly aided by privately and some publicly allotted financial and human resources leveraged by Commonweal, PANNA, and CPR. Still, each biomonitoring sample cost an estimated $200 per sample—an "expensive amount" according to the scientists and researchers in the collaborative. This and other relatively high costs associated with the project limited the number of individuals who could be sampled in the study, but the common motivation and cohesion that developed in the Lindsay Collaborative around the moral imperative to empower the community to stop environmental injustice contributed to the transformations unseen in the Parlier Collaborative.

While this ceding of "state space" (Brenner et al. 2003) offers significant opportunities for advocates to exert their own agency and legitimization strategies, it "privatizes the welfare state" (Marwell 2004) to advocates and their constituencies with potentially great peril. Beginning in the early 1970s, and picking up steam in the mid-1980s under the Reagan/Thatcher consensus, efforts to shrink the size and influence of government to provide publicly funded (and beneficial) supportive services, such as ensuring public health and environmental protection, have made nonprofit and social movement organizations key providers of such services (Brulle 2000; Marwell 2004; Harvey 2006; Pellow 2001, 2007). This development often sets up failures and uneven applications across time, space, and populations, as evidenced in the Parlier and Lindsay collaboratives.

The Lindsay Collaborative also speaks to the problem of subjectivity and identity under neoliberalization (Lietner, Peck, and Sheppard 2007). In particular, while street science is typically associated with collective action and empowered citizenship, it is coupled here with techniques, especially biomonitoring, that tend to individualize environmental hazards and emphasize self-care as a key component of citizenship (Rose 1996). In Lindsay, the perils of the latter are moderated by advocates' efforts to link embodied—and thus individualized—measures of environmental injustice to broader structural factors, including racism, classism, and xenophobia. For example, community advocates repeatedly emphasized how the results of the biomonitoring and pesticide drift catchers pointed to structural contradictions and displacements within California's industrial agricultural model, not merely bodies being in the wrong place at the wrong time.

Conclusion

The "science game" of pesticide regulation continues as street science and regulatory science actors engage each other in political debate. This engagement includes DPR scientists' drawing on the Parlier Collaborative to construct truth claims about pesticide risk from Parlier and truth claims for VOC emissions that meet the standards of the federal Clean Air Act and National Ambient Air Quality Standards; advocates drawing on the Lindsay Collaborative to push for a statewide ban on aerial pesticide spraying near schools; and both sets of actors vying for the public good of scientific legitimacy. It remains to be seen whether such engagement is a limited good subject to zero-sum game rules, or a "comedy of the commons" (Rose 1986) of mutual benefit and enrichment.

This chapter has suggested that both logics are in play. DPR is responding to considerable political pressure from pesticide reform and environmental justice advocates to balance their disinterested scientific standing with the agency's interestedness in the economic development of California's agricultural sector. The agency is also negotiating the stress from the forces of neoliberalization that restrict state action by limiting funding and spaces of legitimacy. Analysis of the Parlier Collaborative suggests DPR's precautionary approach for "reduced-risk" pesticide use in California will continue to privilege voluntary, incentive-based transition programs over command-and-control approaches. Likewise, despite leveraging environmental justice principles and movements for new modes of scientific practice, DPR confronts incentives that encourage applications of regulatory and research science to aid the development of new technologies to maintain California's industrial and high-chemical input form of agricultural production (Matteson, Wihoit, and Robertson 2007).

Meanwhile, pesticide reform and environmental justice advocates are working not only to rehearse their performance of *la realidad*, but are expanding their purview to the everyday practices of the body to transform how California understands and addresses chemical trespass. It is still too early to assess the impact of the precautionary approach of the new pesticide-spray buffer zones and aerial pesticide application rules for the fenceline communities in Tulare County and elsewhere in the San Joaquin Valley. But environmental justice advocates are optimistic that they have provoked a game change by demonstrating the feasibility of innovative participatory scientific research, community capacity building, and precautionary regulatory measures.

Our analysis illustrates the uneven process by which environmental justice movements have transformed different spheres of science and how scientists have resisted and adapted to these changes. We argue that the conceptual clarifications made here between the three spheres of science—regulatory science, research science, and street science—will be useful for exploring future transformations in scientific practice by other social movements. Furthermore, understanding the ways in which scientific modes at once influence each other and resist influence can help inform innovations in environmental justice scholarship and practice. Such attention will help the scientists within advocate, agency, and academic realms address the potential political, economic, and organizational opportunities and constraints on the transformative potential of environmental justice.

Notes

1. The magnitude of pesticide application in the "big three" counties is remarkable even compared to the rest of California, which itself uses a disproportionately high amount of pesticides relative to the rest of the country (Harnly et al. 2005). The latest figures indicate that the San Joaquin Valley accounts for 62.81 percent of the total pounds of pesticides used in California in 2006, while agriculture in the big three counties is also the highest user of pesticides and has maintained that status since at least 2000. These assertions are based on our examination of the California Department of Pesticide Regulation's Pesticide Use Reporting Program data from 2000 to 2006. For more information, see http://www.cdpr.ca.gov/docs/pur/purmain.htm.

2. This conflict "has become an increasingly controversial issue at the urban-agriculture interface," especially in Tulare County and neighboring Kern County since 1999 (Harrison 2006, 507). Some of the impacts of pesticide drift include "serious acute illness (nausea/vomiting, eye/skin irritation, difficulty breathing) and likely contributes to many chronic diseases, including asthma and other lung diseases, cancer, birth defects, immune system suppression, behavioral disorders, and neurological disorders" (Harrison 2006, 507).

3. By "institutionalization," we refer to the process by which established meanings attain a certain character, as embodied in such things as a given program or policy (Jepperson 1991).

4. We do not examine research science because it has not yet entered the political struggles of pesticide reform and environmental justice in a substantive way (see Pinkerton et al. 2010 for an example of some important academic research on the environmental health implications of pesticides in California).

5. The block group demographic data displayed in figure 8.1 come from the 2000 U.S. Census, Summary File 3, tables P087001 and P159H002. County, census-defined "places," and block group boundaries are also from the 2000 U.S. Census.

6. California Government Code §65040.12 (e).

7. Cal/EPA's other agencies—the Air Resources Board, the Department of Toxic Substances Control, the California Integrated Waste Management Board, the Office of Environmental Health Hazard Assessment, and the State Water Resources Control Board—were also assigned various responsibilities, including leading pilot projects relevant to their media of focus within Cal/EPA (see London, Sze, and Liévanos 2008).

8. As a political appointee and agency head, Warmerdam's name is used. Interviewees in civil service positions in DPR and other agencies are provided anonymity.

9. Cumulative impact was defined in terms of looking at a multiplicity of environmental pollutants and their association with public health and multimedia environmental effects, while taking "into account sensitive populations and socio-economic factors, where applicable and to the extent data are available" (Cal/EPA 2005). In addition, *precautionary approach* was defined as "taking

anticipatory action to protect public health or the environment if a reasonable threat of serious harm exists based upon the best available science and other relevant information, even if absolute and undisputed scientific evidence is not available to assess the exact nature and extent of risk" (Cal/EPA 2005).

10. Criteria air pollutants, as classified by the U.S. EPA, pose significant human and environmental health effects, for which acceptable levels of exposure can be determined and for which an ambient air quality standard has been set. Key pollutants are ozone, carbon monoxide, nitrogen dioxide, sulfur dioxide, and PM10 and PM2.5 (CARB 2008).

11. DPR formed an Interagency Work Group in 1997 to explore concerns expressed by community members in Lompoc, California, about the impact of pesticides on community health. As DPR scientists explained to us, the Air Monitoring project in Lompoc "started out adversarial" and did not "flow" as well as the Parlier project did, in terms of community participation and willingness to welcome the monitoring activity in the community. They noted that Parlier was more "neutral." DPR monitored for twenty-seven pesticides in Lompoc and determined that the levels found in the ambient air did not exceed screening levels of concern to human health. Based on such findings, DPR decided to discontinue monitoring activity in Lompoc (CDPR 2003).

12. Monitoring assistance was provided by the California Air Resources Board, the San Joaquin Valley Air Pollution Control District, and the University of California (UC), Kearny Agricultural Center, and the UC Davis Center for Health and the Environment and Western Center for Agricultural Health and Safety.

13. See, for example, the lawsuit *El Comité para el Beinestar de Earlimart v. Helliker*, 416 F. Supp.2d 912 (E.D. Cal. 2006). In it, advocates claimed that DPR's approach to regulating smog-producing pesticides did not meet the stringency required by state and federal law. This suit was won at the Superior Court level, prompting efforts to revise the agency's VOC regulations. However, after the environmental justice advocates lost the case appeal by DPR, such regulatory revisions ceased.

14. *El Quinto Sol*, Spanish for the "Fifth Sun," is derived from Aztec cosmology, and signifies the search for balance in a time of chaos/the end of time. It also represents a reappropriation of "América" into an indigenous/Mexican cultural and political frame.

15. The most recent unsuccessful state-level legislation includes Assembly Bills 1721 (the Health and Safety School Zones Act) and 62. Assembly Bill 1721 would have banned the aerial application of any pesticide within one-quarter mile of schools while in session and when children are present. Assembly Bill 62 would have established a safety zone of no less than 3.3 miles for the aerial application of a pesticide for residential areas and known sensitive sites, such as schools, hospitals, day care centers, senior citizen centers, residential care homes, and farm labor camps. Both bills died in the California Assembly Committee on Agriculture.

16. There are some other notable organizational changes in Commonweal, PANNA, and CPR as a result of their engagement in environmental justice

advocacy. Commonweal—a nonprofit health and environmental research institute in the San Francisco Bay Area—assisted the residents of Diamond, Louisiana, in their relocation efforts away from two petrochemical plants because they found it was the morally right thing to do after visits to the community (Lerner 2005). Commonweal is also known for providing support for community groups through its Biomonitoring Resource Center and an overall "unique synthesis of political advocacy, education, clinical training, social support for cancer victims, and spiritual practices" (Brown 2007, 278). PANNA is a regional branch of the international Pesticide Action Network, and is characterized as having actors networked in a variety of organizational contexts, from the local to the global, engaging a multiplicity of political and economic institutions "depending on which point of access is likely to yield the greatest political payoff" (Pellow 2007, 6–7). PANNA is a founding member of CPR and continues to work closely with the organization in furthering a more local focus on environmental justice in California. CPR has now evolved into a coalition of about 180 members of traditional environmental groups, union locals, children and community health groups, environmental health groups, physician groups, and specifically focused pesticide reform groups. They have all signed under the four-point platform to "eliminate use of the worst pesticides; reduce overall use of the remaining pesticides; promote the use of sustainable pest control solutions in farms, communities, forests, homes and yards; [and] protect people's right-to-know about pesticide use in our neighborhoods, counties, and state" (CPR 2007a).

9

Rupturing Engineering Education: Opportunities for Transforming Expert Identities through Community-Based Projects

Gwen Ottinger

Transformations of scientific activities and approaches have been a central focus of the studies collected in this book. The chapters consider movement toward and barriers to, for example, new ethics for reporting results of biomedical research to study participants (Morello-Frosch et al.) and revised frameworks for assessing the risks posed to minority communities by environmental contamination (Johnson and Ranco; Powell and Powell). But in addition to analyzing shifts in what scientists and engineers *do*, these studies point to shifting relationships between technical practitioners and other participants in environmental justice advocacy. In the process of acting differently—feeding biomonitoring results directly back to community campaigns (Morello-Frosch et al.), or working at a middle level to create effective rural electrification programs (Nieusma)—scientists and engineers disrupt the hierarchies that usually attend technical expertise and, ultimately, redefine the role they play as experts. Indeed, the incomplete transformations and missed opportunities highlighted by several chapters are attributable in no small part to technical professionals' unwillingness or inability to step out of their traditional, "expert" roles: EPA scientists sought Native American participation in risk assessment while retaining all of their power as state experts (Johnson and Ranco); scientists at the Clean Air and Water Network redefined their project to include community organizing and engagement but never found a meaningful way to share their expertise with community members (Hoffman).

Seeing technical practitioners' ability to transform expert roles as central to their ability to take advantage of ruptures in scientific practice begs the question: How do scientists and engineers find—or not find—room to maneuver in their understandings of their professional identities and their appropriate relationships with others involved in environmental justice? Frickel (chapter 1, this volume) points to educational experiences, social justice commitments, and professional networks as relevant factors. In this chapter, I consider the opportunities for redefining expert roles offered by a growing trend in higher education: the use of service learning, increasingly popular across college campuses, to meet calls for reform in engineering education in particular. I argue that environmental justice advocates' calls to transform science and technology demand engineers who can see themselves not as problem solvers, but as participants in problem solving. Toward that end, I show how some engineering educators have been using service-learning projects, in which students engage in community-based projects as a way of mastering course concepts, to critically reflect on their roles as experts in ways compatible with the demands of environmental justice. I then offer the example of "Technology and Environmental Justice," a course that I taught to University of Virginia engineering students. As a central part of the course, students interacted with EJ groups on projects involving website development, data analysis, and/or technical research. Based on my observations of classroom dynamics and student work, I suggest that the projects created opportunities for students to reimagine their professional role as engineers. While students largely continued to see themselves as solution providers, many altered their approaches to solving technical problems, and a few even began to articulate a role for EJ activists and community members as legitimate participants in problem solving. These subtle shifts in student understanding in one case speak to the possibilities of a larger transformation in engineering identity—and to the potency of educational interventions for EJ-inspired transformations of science and technology more generally.

Expert Roles in Environmental Justice: From Orchestra Conductor to Jazz Musician

Calls to transform how scientists and engineers interact with so-called laypeople—or the roles they play as experts—have long been integral to EJ advocacy. Scholars and activists interested in environmental justice have noted several ways in which engineers, scientists, and other technical professionals tend to contribute to environmental *in*justices: by dominating policy processes, by adopting scientific methods that systematically underestimate the effects of pollution on communities, and by designing technologies that both consolidate pollution and reinforce power disparities. In order for technical experts to be contributors to environmental justice instead, many EJ advocates suggest changing the way they interact with community and activist groups, becoming collaborators in research and problem solving rather than simply providers of information and technical solutions.

One way experts exacerbate environmental injustices is through their patterns of participation in public decision-making processes. One of the central tenets of environmental justice is that communities of color, and other communities affected by environmental hazards, should be able to participate "as equal partners" in decisions about siting, enforcement, and risk assessment (People of Color Environmental Leadership Summit 1991), but realizing this principle has been a struggle. Community groups have frequently had to fight for the right to participate fully in local decision-making processes (Cole and Foster 2001; Roberts and Toffolon-Weiss 2001). A number of scholars have also grappled with the problem, theorizing the conditions under which community members could be fairly represented in participatory policy processes (e.g., Guana 1998; Schlosberg 2002). Activists and scholars alike identify the power disparities between community members and the companies they oppose as central obstacles to equal participation; moreover, they note a variety of ways in which experts heighten those disparities. For example, because companies either have or can afford to hire their own experts, they bring to policy processes information to which community members are often denied access (Cole and Foster 2001); this highly technical information is held in higher regard by decision makers than activists' anecdotal or experiential comments (Kuehn 1996). The reliance on experts and their specialized knowledge makes it difficult for activists without comparable knowledge, skills, and access to resources to participate on an even footing with experts (Cable, Mix, and Hastings 2005; Guana 1998), a problem compounded by experts' often patronizing or dismissive attitudes in their dealings with community members (Cole and Foster 2001).

To foster nonexperts' full participation in decision making about technically complex social issues, environmental justice advocates have proposed a number of ways of redefining the relationship between experts and communities. They have suggested, for example, that technical professionals in regulatory agencies and academia actively strive to promote environmental justice through their work (Bailey et al. 1995), that community groups be given the means to hire their own technical experts as part of their participation in policy processes (e.g., Kuehn 1996), and that EJ activists forge collaborations with environmental professionals and academics in order to increase their social capital (Cable, Mix, and Hastings 2005). All of these proposals call for a shift in experts' practice in that they ask them, in one way or another, to incorporate environmental justice into their professional goals. Not all necessarily entail a shift in the basic role of experts—for example, EJ-sensitive regulators would remain providers of technical information and solutions to nonexpert communities—but some, like that of Cable, Mix, and Hastings (2005), do suggest the need for experts to cease to be sole owners of technical problems and instead become collaborators with community groups, a suggestion amplified by those critical of the shape of scientific knowledge itself.

Experts' methods for producing knowledge about the effects of pollution on communities are a second obstacle to environmental justice, in that they often obstruct activists' efforts to see serious health hazards acknowledged and remedied by decision makers. Scientific studies of chemical health effects, for instance, are held to standards of proof that reflect scientists' values of certainty and rigor; these, however, come at the expense of efforts to pursue precautionary policies that would protect communities suspected to be endangered by pollution (Bryant 1995; Head 1995). Little research is done to investigate the environmental conditions in polluted communities—for example, toxicological studies seldom look at the effects of combinations of chemicals, whereas residents are never exposed to just one chemical at a time (Head 1995; Tesh 2000), and air monitoring is not routinely conducted in a way that can adequately characterize exposures. Finally, although "local knowledge," historically absent from scientific studies, is increasingly accepted as an important component of environmental regulatory agencies' health risk assessments (Corburn 2005), nonexperts engaged in collecting their own data about chemical exposures and health effects still struggle to have it acknowledged by decision makers (Allen 2003; Ottinger, 2010).

Recognition of the limitations of scientific knowledge, especially that surrounding environmental health and justice issues, has prompted calls for a new kind of science that, again, implies a new role for experts. In particular, scholars and environmental justice advocates argue for scientific research that actively incorporates the questions and insights of nonscientists; among them, Brian Martin (2006) imagines a "science shaped by a citizen-created world," Alan Irwin (1995) dubs it "citizen science," and Jason Corburn (2005), writing specifically about environmental justice, develops the idea of "street science."

A science shaped by nonscientists, these proposals make plain, requires more than just "lay" participation in knowledge production. It also requires technical professionals to conceive of their roles in the process differently, in order to hear and accommodate the insights of their new scientific collaborators. Corburn (2005, 215) offers a particularly elegant description of how the quest for street science would change technical practice, expressing his ideal through a comparison with jazz:

Jazz music is about improvisation, creativity, and building a group sound. The jazz musician builds on and responds to the actions and tones of other musicians in the group. While the musician reacts in the moment, she does not enter the "jam session" unprepared or without a repertoire of responses. . . . In the *jazz of practice*, professionals will bring their conventional "tool kit" but be rewarded for improvising and being creative. They also will be encouraged to forge new partnerships and to be open to new interpretations of seemingly routine situations.

While jazz is largely about improvisation, accepted rules of procedure exist— when to solo, how to yield to another musician, when to break with tone, and how to "bring back the rhythm." It is untrue that anything goes. In the *jazz of practice*, professionals still will bring their disciplinary methods, but they also must learn and respect the rules and norms—the everyday rhythms—of the cultures and communities with whom they work. . . .

By being playful, improvisational, and open to new "players," the jazz of practice encourages professionals to reinvent and remake existing models of environmental health decision making.

While Corburn's description of a jazzy science is largely focused on how the process and activities of knowledge production would change, it also articulates a new relationship between experts and community members. Rather than being conductors, welcoming (or not) the new instruments of local knowledge into the orchestra, scientists are fellow players, helping to build a "group sound" from within the ensemble. They are responsive, they listen to the rest of the group, they know when to take the lead, and they know when to let others do so. In short, in this vision, as in other visions of citizen-influenced science (e.g., Cable, Mix, and Hastings 2005), technical professionals would see themselves as partners of and collaborators with community groups in producing scientific knowledge—rather than as providers of technical information, solutions, or direction.

Like scientific research, technological design (at least as it is usually practiced) also contributes to environmental injustice. Research on the politics of technology shows that technological artifacts help structure social hierarchies and political possibilities. That is, they can embody specific power relations, support certain political systems at the expense of others, and marginalize groups of people—even when design choices appear to be strictly technical (Nieusma 2004a; Sclove 1993; Winner 1986). As a result, technology is implicated in structuring environmental inequalities. For example, technologies that concentrate energy production in massive power plants or oil refineries, presumably to take advantage of economies of scale, produce a situation where the associated hazards necessarily impact populations that are small relative to those that benefit, perpetuating the unequal distribution of environmental hazards. The technical complexity of these operations, moreover, requires regulators to depend on data produced by the facilities themselves to determine compliance with environmental laws and increases the barriers to citizen participation in regulatory decision making—undermining the EJ goals of equal protection and procedural justice. In contrast, alternative designs attentive to scale or ease of monitoring could minimize the impacts of industrial facilities, facilitate public oversight, and help make a facility's risks more proportional to its benefits.

As in the case of scientific knowledge, creating alternative, environmentally just technologies involves reorienting how experts—especially engineers—see themselves with respect to nonexpert communities. Proponents of alternative design argue for the participation of not only user groups but also citizens in general, in order to ensure that new technologies avoid perpetuating oppressive social structures and promote democratic political arrangements (Nieusma 2004a; Sclove 1993). In the environmental justice context, community-member participation in chemical-plant design might introduce values of decentralization and transparency to the design process alongside engineers' goals of efficiency and reliability. Like "street science," this vision of participatory design implies a shift in technical professionals' roles in the process: engineers cannot merely be providers of solutions but must collaborate with other groups, especially technologically marginalized ones, in order to create appropriate technologies. The "jazz of practice" seems to apply equally well here—creativity and improvisation are certainly hallmarks of participatory design.

Combating the contributions of science, technology, and expertise to environmental injustice, then, requires more than changing political processes, scientific research approaches, and technological design. It simultaneously calls for altering how experts understand their own roles, especially with respect to other, nonexpert groups, in creating science and technology. Rather than being the primary providers of information or solvers of problems for nonexperts, technical professionals working for environmental justice are asked to be partners and collaborators with them. The transformation of expert roles envisioned by EJ advocates has important resonances with the transformations that some engineering educators, including those employing service learning as an educational strategy, are hoping to produce through their curricular reforms.

New Roles for Engineers in Engineering Education Reform and Service Learning

How engineers see themselves in relation to other groups is among the things deemed in need of change by current efforts to reform engineering education. As part of the project of defining goals for engineering education in the twenty-first century, for example, the National Academy of Engineering's (NAE's) report, *The Engineer of 2020: Visions of Engineering in the New Century*, calls on the engineering profession to "build a clear image of the *new roles for engineers*, including as broad-based technology leaders, in the mind of the public and prospective students" (National Academy of Engineering 2004, 5; emphasis added). The need for engineers to assume new roles in society and in the public imagination stems from perceived changes in the nature of engineering. The NAE and others concerned with engineering education argue that globalization, rapid advances in science and technology, and the increasing urgency of environmental problems—all trends that will shape engineering work in the twenty-first century—make cultural, political, and economic factors more central to the work of engineers and at the same time increase the variety of groups with a stake in engineering solutions to complicated, societywide problems (e.g., Abdul-Wahab, Abdulraheem, and Hutchinson 2003; Downey 2005; Splitt 2004; Wulf and Fischer 2002). As a result, ABET, the engineering accreditation board, now requires that engineering students graduate with "an ability to communicate effectively," "an ability to function on multidisciplinary teams," "a knowledge of contemporary issues," and "the broad education necessary to understand the impact of engineering solutions in a global, economic, environmental, and societal context" (ABET 2008, 2), among other "professional" (as opposed to technical) skills (see Shuman, Besterfield-Sacre, and McGourty 2005). With their emphasis on communication, teamwork, and the broad contexts for engineering, these criteria suggest that not only engineers' skills but also how they interact with a variety of stakeholders are at stake in current calls for reform.

Among the ways that engineering educators have proposed to fulfill ABET's "professional skills" requirements is by integrating service-learning projects into the engineering curriculum. As a general educational strategy, service learning involves students in projects for community and nonprofit groups; in the process of helping the community, students master curricular content through supplemental readings, critical classroom discussion, and structured reflection on their experiences (Peterson 2009). Increasingly popular across college campuses since the mid-1980s (Ashburn 2009), service learning began to be institutionalized in engineering programs in 1995, when Purdue University established Engineering Projects in Community Service (EPICS) (Coyle, Jamieson, and Oakes 2005). The growing popularity of service learning as a strategy for engineering education is evidenced by the expansion of EPICS into a national program in 1999, as well as the establishment of the *International Journal for Service Learning in Engineering* in 2005.

Engineering educators who use service learning in their classes argue that community-based learning experiences are valuable because they give students compelling, real-world contexts for engineering design and problem solving, and because they help students develop the "professional skills" specified in the ABET criteria, especially the ability to work on multidisciplinary teams and an awareness of the political, economic, and social issues surrounding engineering work (Coyle, Jamieson, and Oakes 2005; Dewoolkar et al. 2009; Duffy, Tsang, and Lord 2000). Indeed, service learning is regarded as especially valuable by some engineering educators because it allows the ABET's "professional skills" criteria to be addressed *while* students are learning "technical" content, reducing the need to introduce separate courses on the social and ethical contexts of engineering into an already packed curriculum (Duffy, Tsang, and Lord 2000).

Service learning and other educational innovations thus aim to develop students' awareness and understanding of the diverse, apparently non-technical, factors that affect engineering work as part of reformers' calls to prepare engineers to fulfill new roles in society. In the process, however, they tend to situate engineers at the heart of the problem-solving process——a position that has long been central to the identity of American engineers (Downey 2005). Seeing them as ultimately responsible for providing solutions to complex, now *socio*technical problems, most reformist and service-learning projects focus on teaching engineering students to understand and incorporate heterogeneous political and social factors into their problem solving, rather than fundamentally changing their roles in the problem-solving process.

Environmentally oriented projects for engineering education reform, for example, argue that environmental quality and sustainable development are not solely technical issues. At a minimum, they acknowledge the importance of social and political factors in the environmental realm (e.g., Abdul-Wahab, Abdulraheem, and Hutchinson 2003); however, many go further, positing social justice as an important outcome of sustainable engineering (Mihelcic, Phillips, and Watkins 2006) and citing the need to engage stakeholders from diverse perspectives in the process of creating engineering solutions (Bhamidimarri and Butler 1998; McLaughlan 2007; Slater et al. 2007). Similarly, service-learning projects, many of them also explicitly concerned with sustainability (e.g., Goff et al. 2010), emphasize the importance of stakeholder involvement in defining the engineering problems student teams will tackle. Students in Virginia Tech's ROXIE program, for example, spend time volunteering with community partners before identifying a need they think they can fulfill (Goff et al. 2010); EPICS teams meet with their project partners not only to collectively define a project and its goals but to get feedback on their progress throughout the project (Coyle, Jamieson, and Oakes 2005). "Allow[ing] for those with needs to define those needs" is, in fact, identified by engineering educators as a principle of good practice for service learning (Duffy, Tsang, and Lord 2000, 2).

But even projects that directly involve nonexpert stakeholders with the intention of developing engineering students' "professional skills"—thus fostering their ability to fill new roles in a globalizing world—appear to preserve the engineer's place at the heart of the problem-solving process. In particular, sharp distinctions are often drawn, at least conceptually, between engineers' work and its contexts; the latter are then positioned with respect to engineering work as "constraints," "limitations," or "inputs." For example, immersion in rural, developing-world communities is said to make students in a program that combines an engineering master's degree with Peace Corps service "especially aware of societal and community issues that impact the success of an engineering project" (Mihelcic, Phillips, and Watkins 2006, 434), and one student is cited as saying that she "learned how social and economic constraints affect sanitation improvements" (Mihelcic, Phillips, and Watkins 2006, 435). The same separation between technical work and social contexts is also evident in much of the writing about engineering service learning. A University of Vermont team, for instance, describes the goal of their "vertically integrated" service-learning curriculum in this way: "We wanted our students to become capable of considering short- and long-term environmental, social, political, regulatory, and economic issues while identifying, defining, and solving engineering problems" (Dewoolkar et al. 2009, 1257). Casting social issues as forces external to engineers' work allows engineers to maintain their place as solution providers even in interactions with community groups; indeed, students and educators alike often talk about the outcomes of service-learning projects in terms of "providing solutions" and "delivering designs" (see Coyle, Jamieson, and Oakes 2005; Dewoolkar et al. 2009; Goff et al. 2010).

Service learning and other projects for engineering education reform that emphasize the importance of social and political contexts to engineering solutions thus tend to assume that engineers will continue to act as providers of solutions—but that they will come to have a broadened understanding of the problems to be solved. The possibility that their new understandings of the social aspects of engineering problems might require them to adjust their solution-provider identities and interact differently with respect to nonengineers—that is, the kind of change envisioned by environmental justice advocates—is rarely entertained. The work of a few engineering educators, however, suggests how calls for engineering reform generally and service learning in particular might be used to get students to think differently about their roles as experts.

The work of Gary Downey and his colleagues, for example, calls explicitly for a change in engineering identities (Downey 2005; Downey et al. 2006). Downey (2005) argues that, rather than basing their professional distinction on their unique abilities as problem solvers, engineers should begin to see themselves as participants in the process of "problem definition and solution" (PDS). By doing so, they would not only acknowledge the role that they already play in the hybrid process of defining social problems in such a way that they can be solved through technical intervention, they would also *expect* to participate in problem definition, to have to collaborate in the process with "people who define problems differently" (p. 590), to assess the implications of potential solutions for diverse groups, and to mediate among stakeholders.

Downey's proposal that engineering identities move from a problem solving to a PDS orientation does not explicitly address the question of how engineers' interactions with other groups might change in the process. In fact, to a large extent, engineers are still envisioned as being in control of the PDS process: noting that engineers would be responsible for incorporating the visions of all stakeholder groups into their work, Downey (2005, 590) says, for example, that "PDS engineers would be the only participants who expected and were expected by others to explicitly address both the technical and the non-technical dimensions of the processes at the same time." Nonetheless, other aspects of his argument point toward ways that the PDS approach would—or could—

transform how engineers see their functions with respect to those of other participants. Downey refers to the process of problem definition as "collaborative work" (p. 590) among engineers and a variety of nonexperts; moreover, he describes the process of defining problems as one in which participants come to have ownership of the problems. Collaborative problem definition thus suggests the possibility for co-ownership of problem solving by a wide range of groups, not just engineers. Further, in aiming to equip students to work with people who define problems differently than they do, the PDS model not only encourages awareness of multiple perspectives (as most reform projects do), it calls explicitly for engineers to be aware of their own perspective and its limitations, and it expressly aims to cultivate a predisposition among engineers to value the perspectives of others (Downey 2005; Downey et al. 2006).

In their work at Smith College, Riley and Bloomgarden go even further toward fundamentally redefining the role of engineers with respect to other groups. Involving students in a project with a community-based organization, their course on "Engineering and Global Development" was structured expressly to challenge traditional, hierarchical relationships of expertise: students' work with the community, which included the expectation that students would be learning from their community partners, was supported by a classroom experience that rejected "the 'expert' role for the instructor in favor of the 'facilitative' role" (Riley and Bloomgarden 2006, 51). In their assessment of the course, Riley and Bloomgarden in fact lament that they were not better able to disrupt expert roles through the course, suggesting that perhaps the very model of service learning in engineering was to blame:

Ultimately, the model in which engineering students act as engineers on a project to benefit a community client imposes expertise onto students in a way that can undercut some of the pedagogical goals of service learning. . . . Different models may be more effective—for example, ones in which students act as apprentices to community members, so community knowledge is formally recognized. Students can then share what specialized knowledge they have from a more appropriate place—one that facilitates two-way transfers of knowledge. (p. 57)

In identifying the limitations of the course, Riley and Bloomgarden continue to push for ways that service learning might contribute to the transformation of traditional models of expertise, suggesting not only student apprenticeships but longer-term relationships with community partners as possible remedies.

Neither Downey's PDS model nor Riley and Bloomgarden's "liberative pedagogy" approach to service learning has explicitly engaged environmental justice issues. Nonetheless, their aims are consistent with EJ advocates' calls for engineers and other technical professionals to assume different roles with respect to nonexpert community members and other stakeholders in EJ: both projects suggest the possibility of engineers going from solution providers to participants in problem solving. Furthermore, both use engineering educators' interest in developing students' awareness of, and facility with, the field's global contexts as an opportunity to begin shifting the ways that engineers think about their roles.

Engineering Roles in an Undergraduate Course

In "Technology and Environmental Justice," a course I taught at the University of Virginia in 2007 and 2008, I sought to merge environmental justice advocates' calls for reformed models of expertise with engineering educators' interest in equipping students for new kinds of professional roles. Through service-learning projects, undergraduate engineering students enrolled in the course became participants in environmental justice campaigns. They were encouraged to think critically about their roles as "experts" in their interactions with community and nonprofit groups—with the aim that they might also come to see themselves as participants in problem solving, rather than as solution providers.

The class thus served as a modest test of one of this book's central premises, that expert identities can be, and are being, transformed through technical practitioners' participation in the environmental justice movement. I found that students did change as a result of their work on EJ projects—they developed genuine sympathy for, and commitment to, the communities with which they were working, they began to appreciate the gaps and uncertainties entailed in scientific knowledge, and they embraced ways of defining problems contrary to those that their engineering training would have suggested. Pronounced in some students and absent in others, these changes clearly did not amount to a wholesale transformation of engineering identity; in fact, even the most affected students did not cease to see themselves as solution providers. However, the shifts in students' perspectives and attitudes that did occur offer evidence that service-learning projects with EJ groups created productive ruptures in engineering students' understandings of their appropriate roles in interactions with nonengineers—and laid the groundwork for students interested in EJ and other social justice issues to ultimately define their professional roles in ways that depart from traditional solution-provider models.

EJ Projects for Engineering Students

"Technology and Environmental Justice" was offered as part of the Science, Technology, and Society (STS) curriculum at the University of Virginia. All undergraduate students in the School of Engineering and Applied Science must take four STS courses in order to graduate; "Technology and Environmental Justice" was one of a number of courses that they could select to fulfill the requirement for a sophomore-level course devoted to developing their understanding of the interrelationships between technology and society through the in-depth study of a particular topic area. Students who took the course came from across the engineering disciplines and were primarily in the second and third years of their respective programs. While some enrolled because of an interest in environmental issues, others chose the course because it fit well with their schedule. Few had any preexisting knowledge of environmental justice in particular.

The course, taught in spring 2007 and spring 2008, included the study of key concepts establishing the significance of science and technology to environmental justice issues and, less typical for a social science class, term projects that served technical needs identified by environmental justice organizations. Students' conceptual grounding in EJ issues centered on a few main themes. Readings from the environmental justice literature, especially Cole and Foster 2001, were used to establish environmental injustice as a structural problem. Students were then introduced to the idea of the systemic limitations of scientific knowledge and to the concept of local knowledge, particularly in the context of environmental health science, through research at the intersection of environmental justice and science and technology studies (e.g., Corburn 2005; Tesh 2000). The idea that technology itself has consequences for structures of environmental injustice was a third theme of the course; however, because little has been published on the topic, the class collectively applied theories of the politics of artifacts to environmental justice, creating a set of design criteria for environmentally just technology to parallel Sclove's (1993) design criteria for democratic technologies.[1] Finally, students read critiques of, and proposals for, technical professionals' roles in the environmental justice movement.

The development of concepts at the intersection of environmental justice and STS gave students context for the term projects that were a centerpiece of the course. Prior to the start of the semester, environmental justice organizations whose mission includes providing technical support to community groups were asked to identify projects suitable for students with basic quantitative, programming, and research skills.[2] The ideal project, the organizations were told, would be of direct importance to the ongoing campaign of a particular community group[3] but unlikely to be accomplished without students' aid, whether because the organization lacked the technical skills or simply because it was low priority relative to other campaign-related tasks. The projects that they identified included assembling and interpreting data from community-based air-monitoring efforts, upgrading web-based applications for pollution logging and data interpretation, and researching alternatives to polluting industrial processes in use at facilities against which communities were campaigning.

At the beginning of the semester, the class was divided into teams of four to six students, and each team was assigned a project. Groups were expected to work independently to accomplish their projects, and to speak periodically to organizers at the environmental justice organizations that had agreed to help supervise the projects.[4] Class time was set aside to familiarize students with the resources available to them through the University of Virginia's libraries, to check on students' progress on the projects and figure out how to handle any obstacles that they faced, and to speak by phone to representatives of the communities in Louisiana and Ohio on whose behalf the students were working. Students' work on their projects culminated at the end of the semester in written reports summarizing the results of their technical work and analyzing their projects as examples of the limitations of expert knowledge and/or the politics of technology in environmental justice issues. Students also gave public presentations that both reported on their projects and explained their significance to environmental justice issues. Both the written and oral reports were evaluated primarily (though not exclusively) on the quality of students' STS analyses.

Project-Based Transformations

As a result of working on environmental justice projects, many students developed a personal investment in EJ issues. I heard a number of students say in class—and even occasionally to the EJ organizers supervising their projects—that they had begun working on the project simply to earn a good grade, but over the course of the semester, they had come to care about the issues they were working on and the communities that they were working with. Interaction with community members seemed to be crucial to students' sense of commitment: after a slow start on a data analysis project for the community of Chalmette, Louisiana, a group of students in my 2007 class spoke by telephone to Ken Ford, leader of the community group campaigning for environmental improvements at the nearby Exxon refinery. Ford told them about inexplicably high rates of disease in his community, about his own lung cancer, and about local businesses' refusal to host the group's meetings for fear of retribution from the refinery. One member of the student group later recounted that, as they walked out of the conference room where we had had the phone call, a teammate announced, "now it's personal!"; all four of them reported increased motivation and even "passion" for the project, and, indeed, the pace and quality of their work improved dramatically in the subsequent weeks.

In the 2008 class, where students' first contact with community members and activists occurred earlier in the semester, students displayed their sense of commitment to community issues in other ways as well. In telling members of other groups about their projects, many students drew on community members' descriptions of foul odors or widespread illness to situate (and dramatize) the problems that their projects addressed. A few members of groups charged with getting information about processes at facilities so identified with resident-activists that they (perhaps inappropriately) approached their task as something akin to infiltrating the enemy. And many students craved further contact with the communities on whose behalf they were working; one group, developing a website through which Cleveland residents could track the releases of a steel mill, even discussed the possibility of a road trip to Ohio.

Interaction with community activists turned (some) members of the class from grade-motivated students into individuals committed to environmental justice—or at least to a particular environmental justice struggle. At the same time, exposure to community members' perspectives transformed students' understandings of the limitations and uncertainties of scientific knowledge. In class, we discussed reasons that scientific research does not necessarily reveal or result in action on potentially serious health problems that may exist in communities affected by industrial pollution, and students learned about "local knowledge" and how it contrasts with, and complements, scientific knowledge. But these issues were animated by the experiences of community members: students were largely convinced by residents' descriptions of their illnesses and those of their neighbors, yet found that comparing data on chemical exposures—when it was even available—to health guidelines—where they even existed—suggested that pollution was not causing community members' health problems. To reconcile the two stories, students had to improvise, "listening" to newly discovered ways of understanding environmental science along the lines prescribed by Corburn's "jazz of practice." They drew on concepts from the course, letting their projects transform them from abstract concepts into ideas with explanatory power and practical implications. One group, trying to determine why a Norwegian manganese refinery was so much cleaner than its U.S. counterpart, noted and decried the lack of a U.S. standard for manganese levels, in light of known health effects of the chemical. Another group, working with data that documented community exposures to a variety of toxic chemicals but that could not demonstrate that they exceeded standards, went so far as to call the standards "arbitrary"—a claim they justified with reference to work on the uncertainties involved in research on chemical health effects—and suggest that local knowledge needed to be mobilized to better correlate exposure levels with health effects.

Working with environmental justice activists also transformed the way students approached the technical problems with which they were faced, making their problem-solving strategies improvisational rather than formulaic. In part, these transformations stemmed from students' desire to reconcile community members' descriptions of their health problems with scientific accounts that would deny any health effects. The 2007 group whose conversation with Ken Ford gave them new passion for their project first simply compared monitoring data gathered by the community to Louisiana's regulatory standards and found that there were no violations. According to one student, this indicated to the group that there was probably not even a problem, and they began to wonder what the community was complaining about. After becoming galvanized by their conversation with Ford, he said, they asked instead what they could do with the data to help the community, and ended up producing an analysis that still did not show any legal violations, but that demonstrated quite clearly that air quality was significantly worse when the wind was blowing from the refinery—helping to establish the refinery as a culprit in local air-quality issues and call into question regulatory practices that do not consider wind direction in determining violations. Similarly, a group from the 2008 class originally interpreted their task as "dumbing down" the air-monitoring data provided by the Louisiana Department of Environmental Quality (LDEQ) on their website. After speaking with a community member and an environmental justice organizer, they developed a spreadsheet that allowed activists to reinterpret the LDEQ's data in a way that—as they argued in their final report—helps community members correlate their observations with the data and use the data to help explain them.

While many of the changes in students' approaches to problems stemmed from the tension between residents' experiences and official accounts of health effects (or lack thereof), others were a result of students' growing understanding of, and sympathy for, the social and economic realities of communities involved in EJ struggles. The group of students devising a web-based application for logging and mapping pollution complaints around an Ohio steel mill originally bristled at a community organizer's directive that they find a "zero-dollar" solution. By the end of the semester, they had developed a prototype based on freely available software; moreover, they had turned the organizer's observation that most people using the system would be logging in from old, slow computers into a second design constraint central to their mission of creating (in their terms) an "accessible" technology. By letting the knowledge, circumstances, and aims of their community partners inform the design of their projects and their understandings of their tasks, students were starting to imbue their engineering practice with a kind of "jazz."

From Provider to Participant?
Becoming involved in environmental justice campaigns through technical projects resulted in a number of important changes in many students: they developed a sense of commitment to environmental justice, they experienced the ways that scientific uncertainties structure EJ issues, and they altered their approaches to technical problems. What did not appear to change, at least not overtly, was students' understanding of their role with respect to less technically savvy community members and environmental justice organizers. Overwhelmingly, they still positioned themselves as solution providers: reflecting on their roles in their final reports, one group explicitly described their task as providing data and tools for community members; another called themselves "outside consultants"; a third described their role as "experts, informers, and solvers."[5]

Yet there are indications that the transformations that did occur may yet form the basis for students to come to see themselves ultimately as participants in problem solving instead, especially as their professional identities continue to develop. (Midway through their undergraduate education, students were very aware that, although they had more technical skills than activists, they were still less than full-fledged experts.)

While positioning themselves as expert problem solvers in class discussions and final papers, students cited their lack of knowledge about local conditions to argue that activists also needed to be an integral part of problem solving. One student even said that community members needed to be involved because they were the ones who would know what solutions could actually have an impact on local quality of life.

The latter comment in particular suggests that the transformations described above may in fact foreshadow movement by some students toward seeing themselves as participants, alongside EJ activists, in problem solving. Having come to define technical problems differently as a result of interaction with activists, students are less likely to fall back on an easy distinction between the technical and social aspects of a problem—the kind of distinction that underlies many efforts in engineering education to bolster students' awareness of social and political issues in order to make them better providers of technical solutions. Instead, students can understand knowledge of local conditions as integral to the shape that solutions will take, and see activists as integral to the problem-solving process as a result. The experience of contributing to EJ campaigns may have only caused students to begin to see themselves as collaborators (albeit expert ones) in problem solving. Yet by creating the space for them to see themselves as something more than solution providers, the curricular experience laid the groundwork for them to continue reimagining their roles as they become practicing engineers.

Conclusion: The Transformative Potential of Engineering Education

In conjunction with their calls for new ways of doing science, environmental justice advocates have been calling for new kinds of technical experts, ones willing to see themselves not as the source of definitive answers to community environmental issues, but as participants alongside community members in creating knowledge and solving problems. Experts' roles with respect to various stakeholders are also at stake in engineering education reform, and a few engineering educators have used ABET's "professional skills" requirements and universitywide interest in service learning as an opportunity to destabilize engineers' traditional place in problem-solving processes, asking engineering students to collaborate with people who think differently than they do (Downey et al. 2006) and to challenge hierarchies of expertise through their work with community groups (Riley and Bloomgarden 2006). "Technology and Environmental Justice" extended these trends in engineering education with an eye to meeting the calls of environmental justice advocates: by making students participants in environmental justice campaigns, it aimed to encourage students to envision new professional roles for themselves, as participants in problem solving rather than solution providers.

The partial success of the course, like other mixed results reported in this book, is instructive. Students did not cease to see themselves as solution providers—whether due to the way that the service-learning context imposed expert roles on them, insufficient in-class preparation to reflect on those roles, or the cultural durability of the idea of engineer as problem solver (all factors discussed by Riley and Bloomgarden 2006). Yet many students developed far more than an "awareness" of the "social contexts" for engineering, the usual goal of engineering education reformers; rather, they actually came to value local knowledge and see community groups as legitimate participants in problem solving. Participation in environmental justice activities was part of a transformation of students' notions of their role as professional engineers, and it clearly created a rupture in how they saw themselves with respect to "people who think differently," to use Downey et al.'s (2006) term. How these young engineers will maneuver in the space opened up by their EJ experience remains to be seen.

While this case offers lessons for the design of comparable courses—the need for more critical reflection, for example, and more direct involvement with EJ activists—its more significant contribution in the context of this book lies in seeing engineering education as a point of leverage for environmental advocates wishing to replace forms of science, technology, and expertise that produce and exacerbate environmental injustices with more participatory, egalitarian versions that work in the service of EJ. If the transformation of technical practices rests in part on the transformation of expert identities, then policymakers would do well to create incentives for educational experiences that require budding engineers to acknowledge—and hopefully come to value—the contributions that nonengineers can make to solving technical problems, and that give technical professionals the tools to reflect critically on the roles they play with respect to others who rely on their work. And if educational experiences can be an early source of rupture in engineering and science, creating space for reenvisioning technical practices and identities that students may not be able to take advantage of until later in their careers (see Frickel, chapter 1, this volume), then environmental justice advocacy groups should come to regard participation in university-based service-learning projects as a long-term investment, whose value lies not primarily in what a given student group produces for the community but in the quality and nature of expertise likely to be available to EJ advocates in the future.

Acknowledgments

I would like to thank Ben Cohen, Jason Delborne, Francisco Doñez, Scott Frickel, Wyatt Galusky, Karen Hoffman, and Dean Nieusma for their comments on an early version of this chapter. I am also indebted to Donna Riley for her thoughtful feedback, especially her help in situating this work in the larger contexts of service learning and social justice in engineering education. Many thanks, finally, to my two classes of engineering students at the University of Virginia for their spirited engagement with environmental justice issues.

Notes

1. The development of this theme was a major difference between the 2007 and 2008 versions of the course; because of the paucity of literature that speaks specifically to this issue, it was left relatively undeveloped in the 2007 class.

2. In 2007, projects came from the Center for Health, Environment, and Justice and the Louisiana Bucket Brigade (LABB); in 2008, they came from LABB and Ohio Citizen Action.

3. This requirement was added for the 2008 projects, and the discussion in this chapter applies only to students who worked on projects that fit that description (five groups in 2008, one in 2007). In the 2007 version of the course, three student groups did technical research of general interest in the environmental justice movement (e.g., determining the local impacts of supposedly green waste disposal and energy production technologies), while a fourth worked on a project directly associated with a particular community campaign. The difference in the quality of students' experience was marked: students' direct interaction with a community member and a community organizer paved the way for the transformations I describe here, and I did not observe them to any significant extent in the students from the three other 2007 groups.

4. Ordinarily, and ideally, service-learning projects would involve closer contact between students and the community groups with which they worked. My desire to have students work on environmental justice projects in particular, however, required collaboration with groups outside of the region surrounding the University of Virginia, which is relatively rural and devoid of most heavy industry.

5. Student characterization of their roles closely resembles that of other engineering students involved in service-learning projects; see, for example, Goff et al. 2010.

Afterword: Working "Faultlines"

Kim Fortun

Pioneering science studies scholar Sharon Traweek has taught us to forever be on the lookout for intersections where people with different backgrounds, perspectives, skills, and status come together. These intersections are like faultlines, the lines between tectonic plates where earthquakes are likely to happen as the plates move past one another. Traweek's geological metaphor is apt. Faultlines are significant intellectually and politically because they are sites where change is likely to originate, possibly transforming entire landscapes. Focusing analytical attention on "faultlines" is thus a way to understand and even help stimulate change.

The editors and authors of this book have focused their attention in this way, examining the particularly volatile space of environmental injustice, and the intersections there between scientists, engineers, activists, and local communities. What happens there? How does environmental injustice create ruptures in scientific and engineering practice? How do scientists and engineers find their way through these ruptures? What openings become visible as we come to better understand the many kinds of trouble environmental justice creates, and the way it keeps scientists and engineers in motion, forever needing to extend their expertise in innovative ways?

The Trouble with Environmental Injustice

Environmental Justice is the fair treatment and meaningful involvement of all people regardless of race, color, national origin, or income with respect to the development, implementation, and enforcement of environmental laws, regulations, and policies. EPA has this goal for all communities and persons across this Nation. It will be achieved when everyone enjoys the same degree of protection from environmental and health hazards and equal access to the decision-making process to have a healthy environment in which to live, learn, and work.

—U. S. Environmental Protection Agency (2009)

Environmental injustice is often defined as uneven distribution of environmental hazards or burdens. Such a definition is accurate yet does not capture the complexity of what environmental injustice entails. Anyone involved in spaces of environmental injustice quickly learns this. Environmental injustice is troubling, not just because it makes people sick and exacerbates racism and other social stratifications, but also because it confounds conventional ways of thinking about how the world works, and can be changed.

The U.S. Environmental Protection Agency has tried to capture the complexity of the challenge in their definition of environmental justice.[1] First, they highlight fairness, a real conundrum of a concept. Fairness can be said to require freedom from bias and dishonesty, and—for many thinkers (Rawls, for example)—even distribution of goods.[2] Imagining much less realizing "fair treatment" is thus far from straightforward, particularly when it is acknowledged (implicitly in the EPA's definition) that any effort to be fair occurs on uneven ground, shaped by complex histories of bias and exclusion. "Meaningful involvement," regardless of race, color, national origin, or income, in the development, implementation, and enforcement of environmental laws, regulation, and policies, is also quite difficult to define, much less realize—for many, many reasons. As Johnson and Ranco describe in chapter 7, "involvement" almost always happens on terms set by the more powerful party; as Powell and Powell describe in chapter 6, "local" knowledges and interests, ironically, are often judged authentic and legitimate by representatives of the state. Meaningful involvement in environmental law and policy also demands circulation of highly technical information, which is not always accessible even if available, as Delbourne and Galusky describe in chapter 3 and Hoffman in chapter 2.

The political dimensions of environmental justice are very complex, as are the ecological, biological, and medical dimensions. One problem is characterizing the problem. Environmental injustice, by definition, happens to specific groups of people in specific places—places with specific climates, geographies, and histories of land use, beset by weather, local, regional, and global politics, lived in by people with imperfect habits. It can be difficult to isolate causes of sickness and disenfranchisement. It is also difficult to move from theory to ground level and back. Much of the scientific data on toxicity, for example, comes from animal studies. Simple application of findings to a local situation just will not work. Often, too, the "sample size" of those subject to environmental injustices is too small to be of statistical significance to epidemiologists.

And then there is the problem of "information poverty," and the problems with thinking in these terms. There are stunning gaps in what we know about environmental health hazards. Of more than 85,000 chemicals registered for use with the U.S. EPA, for instance, only a fraction have received even minimal toxicological assessment. Even fewer have associated human exposure data (U.S. Government Accountability Office 2009, 2008, 2006, 2005a, 2000).[3]

It is now widely acknowledged that established ways of creating environmental health knowledge—as required by the U.S. Toxic Substances Control Act (TSCA) and carried out by the National Toxicology Program (NTP)—are much too slow, expensive, and noncomprehensive. TSCA is structured such that assessment of toxicity for industrial chemicals (pesticides are dealt with differently) does not have to occur before market introduction unless there is a reason to believe that toxicity is a problem. One does not have to inquire about problems unless one knows enough to think there may be problems. This easily creates a will not to know, particularly in corporate labs, where most toxicologists do their work. Even when toxicity testing is pursued, however, the process is slow and expensive, and there are considerable legal and administrative barriers limiting the ability to act once information does indicate that a chemical is harmful (Schierow 2007). Since TSCA was enacted over thirty years ago, only a handful of substances have been banned under its authority (Pew 2009; U.S. EPA 2007). A key goal of the U.S. National Toxicology Program in the future—toward which exciting steps have been made—is to dramatically accelerate the assessment process, learning to think and regulate in terms of chemical classes, rather than one chemical at a time.

Conflicts over how exposure should be accounted for and represented are also troubling. "Exposure" is actual contact between a contaminant and receptor, whether a human or ecosystem. Emission inventories and satellite data can account for pollution levels at a regional scale but do not say how pollutants trespass the human body. There have been impressive initiatives to build databases of human activity patterns for specific groups—"children," for example—to allow emission data to be connected to individuals, but there are gross generalizations here, too.[4] EPA researchers themselves have long acknowledged that "children" is an insufficiently specific category, and that they need to develop a more nuanced understanding of key age and developmental benchmarks for categorizing children. EPA researchers also acknowledge that available data on children's exposures and activities are insufficient to adequately assess multimedia exposures to environmental contaminants as well as cumulative effect (Cohen Hubal et al. 2000).

Personal biomonitoring, as described by Morello-Frosch and colleagues in chapter 4, skirts some problems with environmental hazard exposure assessment, but is beset by others. Personal biomonitoring detects chemical residues in human blood, tissue, and urine. The number of chemicals for which biomarkers have been developed has increased in recent years but many more chemicals are not yet measurable in this way, and tracking backward from biomonitoring to source is not straightforward, undermining the regulatory and legal purchase. Further, our understanding of how certain body burdens affect health is very limited. Many involved in body-burden studies thus worry about how results should be reported, given that they could stigmatize particular communities without having clinical relevance. Thinking, as Morello-Frosch and colleagues do, in terms of the political as well as clinical efficacy of body-burden information is indeed a paradigm shift, upending long-entrenched modes of risk communication, which many remain committed to— because they are committed to the well-being of study populations.

Entrenched models of risk communication control information flow to mitigate anxiety, especially when there are no obvious ways to treat the problems information reveals. Anxiety is seen as adding to peoples' burden, as "something else they don't deserve." Emergent models of risk communication as embodied in websites such as Scorecard (described in chapter 3 by Delborne and Galusky and in the ethical model put forward by the Morello-Frosch group in chapter 4) leverage and try to channel anxiety into political action, presuming that people "have a right to know." While the established model has been appropriately criticized for serving the interests of polluters, the sincerity of many of its practitioners should not be ignored.

Part of the trouble with environmental justice is that it draws out problems with the way people care about the environment and public health, upsetting or at least destabilizing established moral ideals and practices.[5] Pointing to problems with the way people *do* care is, in many ways, more troubling than pointing to vested interests and the ways people and organizations do not demonstrate care. What it means and looks like to care is called into question.

Thinking in terms of information poverty also suggests problems with the way environmental justice problems are framed. At worst, pointing to information deficits can be a key step in disclaiming the need for precaution and remedial action. Dealing with the paradoxes of information is even more complicated. Biomonitoring, for example, both creates information and draws considerable data gaps into visibility. The simultaneity of having lots of information, and too little, is at issue at many sites of environmental justice, and those involved must learn to work within the contradiction—a contradiction that is easily missed when thinking in terms of information poverty (Fortun 2005). A focus on information poverty also skirts questions about how best to make use of the risk information we do have, and questions about how risk information can be effectively communicated. Too often, it is assumed that "ordinary citizens" are not "technical," discounting both the expertise they do come in with, and also the way risk communication can educate and transform them (Fortun 2009; Powell and Powell, chapter 6, this volume).

Often, what count as reason and "sound science" is at issue. Ironically, environmental injustice gained broad visibility at the very historical moment when complexity became of interest across the sciences. Open systems, nonlinearity, and the systemwide implications of the flap of a butterfly's wing became the rage.[6] Environmental injustice could be said to be the perfect problem, finally ready for explication, explanation, and mitigation. The story is more complicated. Concepts and tools that can accommodate complexity, particularly computer models, have been developed in the environmental sciences and are critical in both scientific and regulatory arenas. Many models aspire to deal with the many variables people living with environmental injustice have long pointed to as significant. But the political backlash has been formidable, even if stunningly stupid in many of its formulations. Peter Huber (1998), renowned for his many tracts on junk science (and related efforts to undermine toxic torts), says the following, for example:

To believe wholeheartedly in micro-environmentalism [concern about toxics], one must be either a savant or put a great deal of trust in savants. In particular, one must put one's trust in computer models. The model is everything. Only the model can say where the dioxin came from, or how it may affect our cellular protein. Only the model will tell us whether our backyard barbeques (collectively, of course) are going to alter rainfall in Rwanda. Only the model can explain why relentless pursuit of the invisible—halogenated hydrocarbons, heavy metals or pesticides—will save birds or cut cancer rates. The cry of the loon is replaced by the hum of the computer.[7]

By Huber's logic, anything that we cannot see "with our own eyes" is not a risk. All prosthetics of seeing, whether technical or otherwise, are disdained.[8] Reason and "sound science" are framed in ways that make environmental injustice impossible to recognize.[9]

How environmental justice problems are framed, delineated, and located historically is enormously significant. Further, most of those involved exhibit the "alreadyness" in the dramatological sense described by Delborne and Galusky. They come to the table encoded by prior experience (with racism, for example) and by tales of other people's experiences (dealing with the EPA at the local level, for instance), occupying at least somewhat prefabricated social roles. This can cause conflict among stakeholders, even when their interests and commitments are fairly aligned. Collaboration within spaces of environmental justice is thus particularly volatile, even at its best.

The trouble with environmental justice is thus deep and complex. The problems at hand frighten, enrage, and perplex. They certainly call for work between people with different backgrounds, status, skills, and perspectives, but make such work especially difficult, and especially admirable.

Scientists and Engineers in Motion

Scientists, engineers, and other kinds of experts are complex and diverse characters, informed by varied intellectual and ethical traditions, shaped by the times and places in which they grew up and work. As Scott Frickel discusses in chapter 1, particular backgrounds, experiences, teachers, and historical events can powerfully shape the motivation and orientation of scientists and engineers. Frickel describes how the involvement of parents in politics set the stage for the work of many of the expert activists he interviewed. He also describes how many in his interview cohort came of age in the 1960s, called to consciousness about many social injustices, as well as to the practical work of responding to gassed student protesters. Like Gwen Ottinger in chapter 9, Frickel also highlights the importance of particular teachers and educational experiences. Some teachers had overt commitments to progressive social change, and passed this mission on to their students. Others involved their students in research that involved the discovery of serious threats to public health, such as the threat posed by thalidomide.

The backgrounds of scientists and engineers certainly mobilize them, in different ways. So, too, does daily work in spaces of environmental injustice. Such spaces are characterized by striking resource constraints (financial as well as informational), strong plays of vested interests, and extraordinarily challenging scientific and technical problems. This book provides many examples, suggesting how scientists and engineers must continually come up with new ways of working, innovating around barriers and stretching their expertise to be responsive to community needs and feedback.

The challenges that scientists and engineers face in spaces of environmental injustice are many. Figuring out how to keep scientific and technical knowledge production "objective" and free of political manipulation is far from straightforward—often demanding deep reflection on how objectivity should be defined. Habitual ways of thinking and talking about science, in particular, suggest that scientists should be disinterested and dispassionate. Scientists working in spaces of environmental injustice demonstrate—and must personally take on—the limits of thinking in these terms, figuring out how they can be motivated and oriented by concern about environmental injustice without compromising scientific rigor and authority.[10]

Defining rigor is also complicated, however. Environmental injustice is not elegant; simple study designs, equations, and models cannot capture it. There are, inevitably, many variables, many of which resist quantification, and much local specificity. Often, there are many pollution point sources, operating alongside many other possible causes of ill health. Causation is not linear. Effects can be delayed, difficult to see, and distributed in seemingly illogical ways. Data are often patchy, noisy, and resistant to interpretation. Conclusions are often tentative and laced with caveats. Rigor is demanded at every stage of the analysis, but its criteria are often emergent rather than established in advance. Narrow definitions of "sound science" are not sufficient.[11] Locale-sensitive field techniques must be innovated and cast in terms that secure authority.[12]

Validating field science has always been particularly challenging, in part because of the cultural authority of laboratory ways of work and knowing. But validating the authority of all kinds of science became particularly challenging in the United States during George W. Bush's administration. Pressures on the environmental sciences were especially apparent and gained some public visibility in coverage of ways scientific reports on global warming were edited by Bush administration officials (Gelbspan 2005). In a 2008 report, the Union of Concerned Scientists documented quite pervasive pressure, affecting scientists and engineers involved in air quality and toxics assessments in particular, both of which are of special concern at sites of environmental injustice (Union of Concerned Scientists 2008).[13] This kind of pressure has put many scientists on guard, but has also taught provocative lessons about how scientific knowledge is produced, communicated, and censored. Censoring sometimes happens at the final report stage, but also at other stages of the research process, from the curricular to the funding stage, through study design, communication of results, and regulatory action.

The science of environmental injustice is complicated, in turn depending on modes of communication that can accommodate diverse knowledge forms and types of claims. Scientists and engineers must be able to communicate with each other, across disciplines, recognizing different ways of handling and thinking about data, precision, robustness, and other signifiers of good science. They also must be able to communicate across the differences between research, regulatory and street scientists that Liévanos, London, and Sze draw out in this book (see their table 8.1).[14] The statutory timetables and legal tests of sufficiency that govern regulatory scientists need to be acknowledged, and it is critical to recognize and validate "street sciences" as valuable even if different, though not entirely different from "establishment" science. As Jason Corburn describes (2005), street science, at its best, makes creative use of multiple traditions, scientific and local. Respecting these differences in work against environmental injustice is challenging but critical. It requires genuine respect for explanatory pluralism (Keller 2001).[15]

Scientists and engineers of all kinds working against environmental injustice need to remain in conversation with policymakers, journalists, activists, and community members. To sustain this, they need to bear in mind the many frustrations activists and community members have faced in their dealing with "the government." With time, activists usually learn which agencies and people within government are allies; anger at "the government," however, is often deep and hard to unsettle—and perplexing if not alienating to people who have enjoyed more privilege. Often environmental justice communities have struggled against narrow constructs of causation embedded in regulatory agency health studies, and endured public hearings that promise an opportunity to participate but fail to deliver. They have also endured suggestions that they don't get it and are hysterical. Scientists and engineers who work with them usually come from very different locations. Many can, however, demonstrate empathy and respect, even if they do not fully understand, and quickly learn the extraordinary pace at which activists and residents dealing with environmental injustice themselves learn, and are able to "meet them halfway." Often, it is a transformative process for all involved, fracturing prefabricated roles and provoking seismic change that begins with a tiny crack.

Openings

Chapters in this book draw out many ways scientists and engineers are needed in spaces of environmental justice, many roles they can play, and many barriers they face in their work. How can this help shape future work by scientists and engineers, environmental activists, community leaders, educators, scholars, and the numerous other types of people concerned with environmental injustice?

Most of the chapters here have overt recommendations. Some have pointed to the need to change laws—such as TSCA—such that the burden of proof for product safety is borne by the industries that produce and profit from these products.[16] The Morello-Frosch group calls for changes in the way health-study results are circulated, in ways that call on us to rethink the uses of health information (beyond the clinic, in particular) and also the capacity of "ordinary citizens" to deal with information that is complex, riveted with uncertainty and threatening in, a number of ways (implicating not only health, but also access to jobs and insurance, real estate prices, and potential for investment in neighborhoods and even regions). Nieusma, like Johnson and Ranco, calls on us to rethink the kinds of expertise that scientists and engineers need to have, highlighting the importance of them being able to really listen to knowledges different from their own, understanding what Nieusma terms "hierarchies of knowledge." Nieusma also shows how scientists and engineers need to have elicitation skills: just listening is not good enough. They need to be able to help the people they work with imagine and articulate viable, sustainable, and just futures. Ottinger shows how such refiguring of expertise can start in the university classroom, if pedagogy is fundamentally rethought, involving science and engineering students in environmental justice controversies so that that the development of their expertise is oriented toward concrete problems, technical as well as social.

The stakes are high. What is called for is change of the most fundamental sort.

This should not be surprising, given the focus here on "faultlines" and "ruptures" where experts of different kinds come together and work through complicated and stubborn problems that arise from and depend for resolution on entwined biophysical, technological, political economic, social, and cultural systems.

Traweek argues that focusing on "faultlines" can help us draw out ways "knowledge is being defined and made at the edge of times and places called modernity" (Traweek 2000, 48). By this, she points to ways that deeply entrenched ideas about rigorous practice and right knowledge begin to tremble and shift at faultlines, toppling and cracking structures that once seemed unmovable, sometime revealing things that were previously hidden. The ground is literally mined under, and something new must be built. Contributors to this book show us how and why this is happening in spaces of environmental injustice.

Grassroots environmental activist Diane Wilson puts the challenge and opportunity in her own terms in her book about toxic dumping in Calhoun County, Texas, titled *An Unreasonable Woman*. Wilson has been called many bad names by those whose actions she has opposed. In casting herself as unreasonable, she is not accepting the pejoratives. Instead, she points to the way work to lessen environmental harms and the power differentials that sustain them must work not only against but also outside convention. Business cannot go on as usual. What once seemed unreasonable must be attempted. New kinds of knowledge must be produced and legitimated; new collaborations must be sustained; problem solving must be intensely innovative. Transformation of many kinds, as the editors of this book highlight, is what environmental justice both drives and demands. Scientists and engineers, environmental and community activists, educators and scholars should ready themselves for the seismic shifts already underway.

Notes

1. The U.S. Government Accountability Office has criticized EPA's implementation of environmental justice initiatives for reasons of interest here. Focusing on clean-air rulemaking in a 2007 report (which follows a 2005 report), GAO notes that while workgroups involved in rulemaking were told to consider environmental justice issues, they were not provided training to prepare them to do so, "limiting their ability to analyze such issues." GAO also criticized EPA for not providing a clear rationale and supporting evidence for final decisions on environmental justice–related matters. GAO thus criticized EPA for insufficient treatment of environmental justice as a matter of expert concern, with expertise understood as a cultivated capacity for analysis. Understood in this way, there are, of course, many types of expertise, within and beyond science and engineering (GAO 2007, 2005).

2. In building, fairness is opposed to intersections and faultlines. To make something fair, is to bring things (rivet holes, for example) into perfect alignment, or to smooth over junctions. Traweek argues that critical intellectual possibilities are missed when aesthetic bias against faults keeps us from attending to them. She explains that having spent so much time in California and Japan, well-known

"fault zones," she finds herself anxious when in areas like New York or Boston "where people did not realize they were living on or near active fault lines and have made no public or private preparations for earthquakes." This made her wonder "why in some places we attend to our faults and in other places do not." Thus invested in the double entendre of "faults," Traweek describes her "Fault-lines" essay as about how to examine faults in the ecologies of our minds. By implication, Traweek argues that attending to faults is critical to the pursuit of (political) fairness.

3. The 2000 U.S. Government Accountability Office report says that of 579 chemicals reported in the U.S. Toxic Release Inventory (which only records *legal* emissions), only 50 had associated human exposure data. Of 243 pesticides listed by the EPA and the U.S. Department of Agriculture as of "potential concern," only 32 had associated human exposure data. Of 168 chemicals on the EPA's list of air toxics, only 27 had associated human exposure data (U.S. Government Accountability Office 2000, 14).

4. See the EPA's Consolidated Human Activity Database (CHAD), for example (http://www.epa.gov/chadnet1/).

5. Timothy Luke and Arun Agrawal (2005, 162) show how neoliberal ideals are built into many environmental projects today, producing problematic "environmentalities." Agrawal describes how environmental projects often include explicit programs to produce new environmental subjects, neoliberalizing the way people care about the environment.

6. Traweek (2002, 27) describes the shift as follows: "In the sciences the dominant conceptual strategy during the modernist period was to develop theories and design experiments that investigated or displayed simplicity, stabilities, uniformities, taxonomies, regularities, hierarchies and binaries. During the 1960s these aesthetic criteria begin to be displaced by another set; increasingly, researchers in the sciences preferred to investigate complexity, instability, variation, transformations, irregularities, diverse organizational patterns, and spectra and a new set of tools were developed to help them conduct those searches." Describing this shift in the social sciences, Traweek (2002 27, 31) notes that earlier (modernist) investments were in what did not change, and in what was not affected by local circumstances (historical, political, or economic): "The newer questions ask about the specific conditions in which our various ways of making sense develop; it is now assumed that those conditions must be known before we can successfully identify any sort of patterns."

7. http://www.manhattan-institute.org/html/_commentary-saving_the_environ. htm (accessed October 11, 2010).

8. Paul Edwards (1992, 2010) has explicated the politics of computational ways of knowing in climate science.

9. Scholars of postcolonialism have powerfully drawn out problems and politics of recognition. See, for example, Povinelli 2002.

10. Kelly Moore describes how politically committed scientists in the postwar period worked to "unbind" their profession and its cultural authority, choosing not to engage in the boundary work necessary to sustain such authority. The

movement of scientists and engineers within spaces opened up by environmental justice seems to me different. While usually aware of problems with the way scientific authority can be exercised, scientists and engineers working against environmental injustice nonetheless work to secure scientific authority, working extremely hard to convince themselves and others that their methods, data, and results are valid. They engage in boundary work more to validate new methods and claims than to uphold established methods and claims—though rigorous and creative use of available methods and tools of all kinds is usually key, given both resource constraints and awareness of the need to validate findings in terms sensible to people who remain wedded to old ideals.

11. Calls for "sound science" are not straightforward. Writing about tobacco science, Ong and Glantz (2001, 1749) explain that "public health professionals need to be aware that the 'sound science' movement is not an indigenous effort from within the profession to improve the quality of scientific discourse, but reflects sophisticated public relations campaigns controlled by industry executives and lawyers whose aim is to manipulate the standards of scientific proof to serve the corporate interests of their clients." Also see Michaels 2008.

12. The particular challenge of validating field science (as opposed to laboratory science) is explicated by Robert Kohler (2002). Kohler describes how the "new natural history" first attempted to orient and validate itself by incorporating laboratory techniques into its practice. This approach resulted in important theoretical advances, and a number of practitioners were wedded to it. Ultimately however, according to Kohler, treating the field like a laboratory failed, and the most exciting advances occurred once this was acknowledged. Between the 1930s and the 1950s, as naturalists took on the challenge of developing techniques particularly suited to the field context in which they worked, ecology and evolutionary taxonomy built identities of their own as respectable scientific disciplines. Chris Sellers (1997) tells a different kind of story in his account of the consolidation of occupational health as a field, describing how the field consolidated and built legitimacy by adopting laboratory approaches and standards, despite the field conditions in which they worked (in the early twentieth century).

13. The disregard for science by the Bush administration implicates STS in ways that are still being worked out. Over the last decade, as the political establishment has undermined science in myriad ways, routine activist and STS critiques of science for serving elite, capitalist, military agendas have lost both explanatory and political purchase. Portraying the failures of science remains important, but is not sufficient. The challenge now is to critically analyze scientific practice in order to support and sustain it. Collaboration rather than contempt is the demand of the day. Focusing on science and engineering within zones of environmental justice thus provides rich opportunities for rethinking how science and engineering can be critically engaged, in ways attuned both to broad political contexts and to what the sciences have become. Locating and drawing out the work of exceptional individuals, as Moore has done for the postwar period, will be important, though a readiness to accept and pursue the political implications of their work seems quite pervasive among scientists and engineers today, particularly in environmental fields. As Traweek highlights in her "Faultlines" essay,

and Cohen and Ottinger note in their introduction to this book, the cultures and orientations of science shift such that what was once exceptional becomes routine. A "civic science" sensibility can thus be seen more routinely today, though shifts in the orientation of the sciences since the 1990s are complicated. It is also the age of the commercialization of biology, for example, and an age during which industry in general has enjoyed enormous if not unprecedented privilege, drawing many experts into its ambit. David Hess (2007, 2009) details and stresses the way technical experts have been influenced by industry in recent years. I nonetheless think it can be argued that the time is ripe for what Mike Fortun calls "friendship with the sciences" (M. Fortun 2005; K. Fortun and M. Fortun 2005).

14. It is also useful to acknowledge differences among research scientists, noting how EPA's culture and research budget compares with the culture and budget at the National Institute of Environmental Health Sciences, for example. EPA's budget is much smaller, and NIEHS's budget is much smaller than those of many other NIH Institutes, including the National Cancer Institute, which does not prioritize the environmental determinants of cancer.

15. Historian of science Evelyn Fox Keller (2002) advances the concept of "explanatory pluralism." In concluding her book, she explains that "the central concern of this book has been with the de facto multiplicity of explanatory styles in scientific practice, reflecting the manifest diversity of epistemological goals which researchers bring to their task. But I also want to argue that the investigation of processes as inherently complex as biological development may in fact require such diversity. Explanatory pluralism, I suggest, is now not simply a reflection of differences in epistemological cultures but a positive virtue in itself, representing our best chance of coming to terms with the world around us" (p. 300). The chapters in this book clearly demonstrate the importance of explanatory pluralism in addressing environmental injustice.

16. On February 26, 2009, the Subcommittee on Commerce, Trade, and Consumer Protection of the House Energy and Commerce Committee held a hearing titled "Revisiting the Toxic Substances Control Act of 1976." There was wide agreement that TSCA reform is called for, and a number of scientists testified in support of reform. Industry representatives acknowledged that TSCA needs to be "modernized." In April 2010, Senator Frank R. Lautenberg (D-NJ), chairman of the Senate Subcommittee on Environmental Health, introduced the Safe Chemicals Act of 2010. Simultaneously, Representatives Henry Waxman (D-Calif.), chairman of the House Energy and Commerce Committee, and Bobby Rush (D-Ill.), chairman of the Trade and Consumer Protection Subcommittee, released a "discussion draft" of a similar measure called the Toxic Chemicals Safety Act. Chemical industry lobbyists have subsequently spent tens of millions of dollars opposing TSCA reform. As this book was going to press, the reforms' success appeared doubtful (Kaplan 2010).

References

Abdul-Wahab, S. A., M. Y. Abdulraheem, and M. Hutchinson. 2003. The need for inclusion of environmental education in undergraduate engineering curricula. *International Journal of Sustainability in Higher Education* 4 (2): 126–137.

ABET. 2008. *Criteria for Accrediting Engineering Programs.* Baltimore: ABET.

Agrawal, Arun. 2005. Environmentality: Community, intimate government, and the making of environmental subjects in Kumaon, India. *Current Anthropology* 46 (2): 161–190.

Agyeman, Julian. 2005. *Sustainable Communities and the Challenge of Environmental Justice.* New York: New York University Press.

Allen, Barbara L. 1998–1999. Women scientists and feminist methodologies in Louisiana's chemical corridor. *Michigan Feminist Studies* 13:89–110.

Allen, Barbara L. 2000. The popular geography of illness in the industrial corridor. In Craig E. Colten, ed., *Transforming New Orleans and Its Environs: Centuries of Change.* Pittsburgh: University of Pittsburgh Press.

Allen, Barbara L. 2003. *Uneasy Alchemy: Citizens and Experts in Louisiana's Chemical Corridor Disputes.* Cambridge, MA: MIT Press.

Allen, Barbara L. 2004. Shifting boundary work: Issues and tensions in environmental health science in the case of Grand Bois, Louisiana. *Science as Culture* 13:429–448.

Altman, Rebecca, Rachel Morello-Frosch, Julia Brody, Ruthann Rudel, Phil Brown, and Mara Averick. 2008. Pollution comes home and gets personal: Women's experience of household chemical exposure. *Journal of Health and Social Behavior* 49 (4):417–435.

American Lung Association. 2007. *American Lung Association State of the Air 2007.* New York: American Lung Association.

Arendt, Maryse. 2008. Communicating human biomonitoring results to ensure policy coherence with public health recommendations: Analysing breastmilk whilst protecting, promoting and supporting breastfeeding. *Environmental Health* 7 (Suppl. 1): S6.

Arquette, Mary, Maxine Cole, Katsi Cook, Brenda LaFrance, Margaret Peters, James Ransom, Elvera Sargent, Vivian Smoke, and Arlene Stairs. 2002. Holistic risk-based environmental decision making: A Native perspective. *Environmental Health Perspectives* 110:259–264.

Ashburn, E. 2009. College makes new connections with service-learning program. *Chronicle of Higher Education*, February 27, 2009, A25–A26.

Associated Press. 1990. Mercury level still high in lake sediment. *Wisconsin State Journal*, May 8, 1990, 3D.

Bailey, Conner, Kelly Alley, Charles E. Faupel, and Cathy Solheim. 1995. Environmental justice and the professional. In Bunyan Bryant, ed., *Environmental Justice: Issues, Policies, and Solutions*, 35–44. Washington, DC: Island Press.

Balousek, Marv. 2000. State of the water: In dry weather, pollutants take highest toll on small streams, report says. *Wisconsin State Journal*, February 8, 2000, 1A.

Bates, Michael. N., Sherry. G. Selevan, Susan M. Ellerbee and Lawrence M. Gartner et al. 2002. Reporting needs for studies of environmental chemicals in human milk. *Journal of Toxicology and Environmental Health, Part A*, 65 (22): 1867–1879.

Beck, Ulrich. 1992. *Risk Society: Towards a New Modernity*. London: Sage.

Beehler, Gregory P., Bridget M. McGuinness, and John E. Vena. 2003. Characterizing Latino anglers' environmental risk perceptions, sport fish consumption and advisory awareness. *Medical Anthropology Quarterly* 17 (1): 99–116.

Bennett, Susan. 1990. *Theatre Audiences: A Theory of Production and Reception*. London: Routledge.

Bhamidimarri, R., and K. Butler. 1998. Environmental engineering education at the millennium: An integrated approach. *Water Science and Technology* 38 (11): 311–314.

Bhatia, Rajiv, Barbara Brenner, Brenda Salgado, Bhavana Shumasunder, and Swati Prakash 2005. Biomonitoring: What communities must know. *Race, Poverty and Environment* 11 (2): 56.

Bijker, Wiebe, Thomas Hughes, and Trevor Pinch, eds. 1987. *The Social Construction of Technological Systems*. Cambridge, MA: MIT Press.

Bishop, R. 1994. Initiating empowering research? *New Zealand Journal of Educational Studies* 29:175–188.

Blum, Elizabeth D. 2008. *Love Canal Revisited: Race, Class, and Gender in Environmental Activism*. Lawrence: University Press of Kansas.

Boal, Augusto. 1985. *Theatre of the Oppressed*. New York: Theatre Communications Group.

Bocking, Stephen. 2004. *Nature's Experts: Science, Politics, and the Environment*. New Brunswick, NJ: Rutgers University Press.

Body Burden Work Group and Commonweal Biomonitoring Resource Center. 2007. Is it in us: Toxic trespass, regulatory failure & opportunities for action. http://isitinus.org/project.php.

Bollier, David. 1991. *Citizen Action and Other Big Ideas: A History of Ralph Nader and the Modern Consumer Movement.* Washington, DC: Center for Study of Responsive Law.

Boston Consensus Conference. 2006. Measuring chemicals in people—What would you say? A Boston Consensus Conference on Biomonitoring. Boston. http://www.biomonitoring06.org.

Bourdieu, Pierre. 1977. *Outline of a Theory of Practice.* New York: Cambridge University Press.

Bourdieu, Pierre. 1990. *The Logic of Practice.* Trans. Richard Nice. Stanford, CA: Stanford University Press.

Brenner, Neil, Bob Jessop, Martin Jones, and Gordon Macleod, eds. 2003. *State/Space: A Reader.* Malden, MA: Blackwell.

Brody, Julia G., Rachel Morello-Frosch, Phil Brown, Ruthann Rudel, Rebecca Gasior Altman, Margaret Frye, Cheryl A. Osimo, Carla Perez and Liesel M. Seryak. 2007. Improving disclosure and consent: "Is it safe?": New ethics for reporting personal exposures to environmental chemicals. *American Journal of Public Health* 97 (9): 1547–1554.

Brown, Phil. 1992. Popular epidemiology and toxic waste contamination: Lay and professional ways of knowing. *Journal of Health and Social Behavior* 33:267–281.

Brown, Phil. 1997. Popular epidemiology revisited. *Current Sociology* 45: 137–156.

Brown, Phil. 2007. *Toxic Exposures: Contested Illness and the Environmental Health Movement.* New York: Columbia University Press.

Brown, Phil, Sabrina McCormick, Brian Mayer, Stephen Zavestoski, Rachel Morello-Frosch, Rebecca Gasior Altman, and Laura Senier. 2006. "A lab of our own": Environmental causation of breast cancer and challenges to the dominant epidemiological paradigm. *Science Technology and Human Values* 31 (5): 499–536.

Brown, Phil, and Richard Clapp. 2002. Looking back on Love Canal. *Public Health Reports* 117:95–117.

Brown, Phil, Steve Kroll-Smith, and Valerie J. Gunter. 2000. Knowledge, citizens, and organizations: An overview of environments, disease, and social conflict. In S. Kroll-Smith, P. Brown, and V. J. Gunter, eds., *Illness and the Environment: A Reader in Contested Medicine,* 9–25. New York: New York University Press.

Brown, Phil, and Edwin J. Mikkelson. 1990. *No Safe Place: Toxic Waste, Leukemia, and Community Action.* Berkeley: University of California Press.

Browne, Janet. 2003. Charles Darwin as a celebrity. *Science in Context* 16: 175–194.

Brulle, Robert J. 2000. *Agency, Democracy, and Nature: The U.S. Environmental Movement from a Critical Theory Perspective.* Cambridge, MA: MIT Press.

Brulle, Robert J., and Jonathan Essoka. 2005. Whose environmental justice? An analysis of the governance structure of environmental justice organizations in the United States. In David Pellow and Robert Brulle, eds., *Power, Justice, and the Environment: A Critical Appraisal of the Environmental Justice Movement*, 205–218. Cambridge, MA: MIT Press.

Brulle, Robert J., and David N. Pellow. 2006. Environmental justice: Human health and environmental inequalities. *Annual Review of Public Health* 27:103–124.

Bryant, Bunyan. 1995. Issues and potential policies and solutions for environmental justice: An overview. In Bunyan Bryant, ed., *Environmental Justice: Issues, Policies, and Solutions*, 8–34. Washington, DC: Island Press.

Buck, Germaine, John Vena, Enrique Schisterman, Jacek Dmochowski, Pauline Mendola, Lowell Sever, Edward Fitzgerald, Paul Kostyniak, Hebe Greizerstein, and James Olson. 2000. Parental consumption of contaminated sport fish from Lake Ontario and predicted fecundability. *Epidemiology (Cambridge, MA)* 11 (4): 388–393.

Bullard, Robert. 1993. Anatomy of environmental racism and the environmental justice movement. In Robert Bullard, ed., *Confronting Environmental Racism*, 15–40. Boston: South End Press.

Bullard, Robert D. 2000. *Dumping in Dixie: Race, Class, and Environmental Quality*. 3rd ed. Boulder, CO: Westview Press.

Burger, Joanna. 2000. Consumption advisories and compliance: The fishing public and the deamplification of risk. *Journal of Environmental Planning and Management* 43 (4): 471–488.

Burger, Joanna, and Michael Gochfeld. 2006. A framework and information needs for the management of risks from consumption of self-caught fish. *Environmental Research* 101 (2): 275–285.

Burgin, Aaron. 2006. Group protests pesticide drift. *Porterville Recorder*, July 19., 1A–5A. http://www.pesticidereform.org/article.php?id=280

Bury, M. 2004. Researching patient-professional interactions. *Journal of Health Services Research & Policy* 9 (Suppl. 1): 48–54.

Cable, Sherry, Tamara Mix, and Donald Hastings. 2005. Mission impossible? Environmental justice activists' collaborations with professional environmentalists and with academics. In D. N. Pellow and R. J. Brulle, eds., *Power, Justice, and the Environment*, 55–75. Cambridge, MA: MIT Press.

Cal/EPA. 2004. *Environmental Justice Action Plan*. California Environmental Protection Agency. http://www.calepa.ca.gov/EnvJustice/ActionPlan.

Cal/EPA. 2005. *Cal/EPA EJ Action Plan Pilot Projects: Addressing Cumulative Impacts and Precautionary Approach, March 25, 2005*. California Environmental Protection Agency. http://www.calepa.ca.gov/EnvJustice/ActionPlan/PhaseI/March2005/CI_PA.pdf.

CARB. 2005. *Environmental Justice Pilot Project: Project Objectives, Pesticide, and Community for Monitoring*. California Department of Pesticide Regulation. http://www.cdpr.ca.gov/docs/envjust/pilot_proj/ej_candidate_discussion_final.pdf.

CARB. 2007a. Cal/EPA Environmental Justice Action Plan: DPR Parlier Project Update, May 31, 2007. California Environmental Protection Agency. http://www.calepa.ca.gov/EnvJustice/Documents/2007/DPRParlier.pdf.

CARB. 2007b. California Department of Pesticide Regulation home page. California Department of Pesticide Regulation. http://www.cdpr.ca.gov.

CARB. 2008. *Consumer Information: Glossary of Air Pollution Terms.* California Air Resources Board. http://www.arb.ca.gov/html/gloss.htm#C.

Carson, Cathryn. 2003. Objectivity and the scientist: Heisenberg rethinks. *Science in Context* 16:243–269.

CDC. 1999. *The First National Report on Human Exposure to Environmental Chemicals.* Atlanta: Centers for Disease Control and Prevention, National Center for Environmental Health.

CDC. 2003. *Second National Report on Human Exposure to Environmental Chemicals.* Atlanta: Centers for Disease Control and Prevention, National Center for Environmental Health.

CDC. 2005. *Third National Report on Human Exposure to Environmental Chemicals.* Atlanta: Centers for Disease Control and Prevention, National Center for Environmental Health.

CDC. 2009. *Fourth National Report on Human Exposure to Environmental Chemicals.* Atlanta: Centers for Disease Control and Prevention, National Center for Environmental Health.

CDPR. 2003. Meeting summary of the Lompoc Interagency Work Group. California Department of Pesticide Regulation. http://www.cdpr.ca.gov/docs/specproj/lompoc/summary041003.pdf.

CDPR. 2005. Environmental justice pilot project: Project objectives, pesticides, and community for monitoring. California Department of Pesticide Regulation. http://www.cdpr.ca.gov/docs/envjust/pilot_proj/ej_candidate_discussion_final.pdf.

CPR. 2007a. Californians for Pesticide Reform: Mission. Californians for Pesticide Reform. http://www.pesticidereform.org/article.php?list=type&type=17

CPR. 2007b. Airborne poisons: Pesticides in our air and in our bodies. Californians for Pesticide Reform. http://www.pesticidereform.org/downloads/Biodrift-Summary-Eng.pdf

Choy, Timothy. 2005. Articulated knowledges: Environmental forms after universality's demise. *American Anthropologist* 107 (1): 5–18.

Clinton, William J. 1994. Federal actions to address environmental justice in minority populations and low-income populations. Executive Order 12898. 59 FR 7629, February 16. http://www.epa.gov/history/topics/justice/02.htm

CNN. 2007. Many carry chemical "burden": The results of body-burden testing in two children are cause for parents' alarm. *Planet in Peril: Environmental Coverage–Special Reports from CNN.com.*

Cohen-Hubal, Elaine A., Linda S. Sheldon, Janet M. Burke, Thomas R. McCurdy, Maurice R. Berry, Marc L. Rigas, Valerie G. Zartarian, and Natalie C. G.

Freeman. 2000. Children's exposure assessment: A review of factors influencing children's exposure, and the data available to characterize and assess that exposure. *Environmental Health Perspectives* 108:475–486.

Colborn, Theo, Dianne Dumanoski, and John Peterson Myers. 1996. *Our Stolen Future*. New York: Penguin.

Cole, Luke, and Sheila Foster. 2001. *From the Ground Up: Environmental Racism and the Rise of the Environmental Justice Movement*. New York: New York University Press.

Collins, H. M., and Robert Evans. 2002. The third wave of science studies: Studies of expertise and experience. *Social Studies of Science* 32:235–296.

Commission for Racial Justice. 1987. *Toxic Wastes and Race in the United States: A National Report on the Racial and Socioeconomic Characteristics of Communities with Hazardous Waste Sites*. New York: United Church of Christ.

Cone, M. 2005. *Silent Snow*. New York: Grove Press.

Corburn, Jason. 2002. Combining community-based research and local knowledge to confront asthma and subsistence-fishing hazards in Greenpoint/Williamsburg, Brooklyn, New York. *Environmental Health Perspectives* 110, no. 2 (Suppl.): 241–248.

Corburn, Jason. 2005. *Street Science: Community Knowledge and Environmental Health Justice*. Cambridge, MA: MIT Press.

Couch, Stephen R., and Steve Kroll-Smith. 1997. Environmental movements and expert knowledge: Evidence for a new populism. *International Journal of Contemporary Sociology* 34:185–210.

Coyle, E. J., L. H. Jamieson, and W. C. Oakes. 2005. EPICS: Engineering Projects in Community Service. *International Journal of Engineering Education* 21 (1): 139–150.

CPR. 2007a. Airborne poisons: Pesticides in our air and in our bodies. Californians for Pesticide Reform. http://www.pesticidereform.org/downloads/Biodrift-Summary-Eng.pdf.

CPR. 2007b. Californians for Pesticide Reform: Mission. Californians for Pesticide Reform. http://www.pesticidereform.org/article.php?list=type&type=17.

Cullen, Sandy. 2008. Signs urged for fish warnings: The Madison Environmental Justice Organization is worried that anglers might not know the health risks of catch they pull from Madison's lakes. *Wisconsin State Journal*, August 19.

Daston, Lorraine, and Peter Galison. 2007. *Objectivity*. New York: Zone Books.

Daston, Lorraine, and H. Otto Sibum. 2003. Introduction: Scientific personae and their histories. *Science in Context* 16:1–8.

Daughton, Christian G. 2001. "Emerging" pollutants, and communicating the science of environmental chemistry and mass spectrometry—pharmaceuticals in the environment. *Journal of the American Society for Mass Spectrometry* 12 (10): 1067–1076.

Davis, D. L., and P. Webster. 2002. The social context of science: Cancer and the environment. *Annals of the American Academy of Political and Social Science* 584:13.

De Certeau, Michel. 1984. *The Practice of Everyday Life.* Berkeley: University of California Press.

Deck, W., and T. Kosatsky. 1999. Communicating their individual results to participants in an environmental exposure study: Insights from clinical ethics. *Environmental Research* 80 (2, pt. 2): S223–S229.

Delborne, Jason A. 2008. Transgenes and transgressions: Scientific dissent as heterogeneous practice. *Social Studies of Science* 38 (4): 509–541.

Delborne, Jason. 2011. Constructing audiences in scientific controversy. *Social Epistemology* 25 (1): 67–95.

Deleuze, Gilles, and Felix Guattari. 1987. *A Thousand Plateaus: Capitalism and Schizophrenia.* Minneapolis: University of Minnesota Press.

Dewoolkar, M. M., L. George, N. J. Hayden, and D. M. Rizzo. 2009. Vertical integration of service-learning into civil and environmental engineering curricula. *International Journal of Engineering Education* 25 (6): 1257–1269.

Di Chiro, Giovanna. 1997. Local actions, global visions: Remaking environmental expertise. *Frontiers* 18 (2): 203–232.

Di Chiro, Giovanna. 1998. Environmental justice from the grassroots: Reflections on history, gender, and expertise. In Daniel Faber, ed., *The Struggle for Ecological Democracy: Environmental Justice Movements in the United States,* 104–136. New York: Guilford Press.

Doege, David. 2006. Hate crime stands against firefighters: They're charged in confrontation with black man. *Milwaukee Journal Sentinel,* November 29.

Downey, G. L. 2005. Are engineers losing control of technology? From "problem solving" to "problem definition and solution" in engineering education. *Chemical Engineering Research and Design* 83 (A6): 583–595.

Downey, G. L., J. C. Lucena, B. M. Moskal, R. Parkhurst, T. Bigley, C. Hayes, B. K. Jesiek, L. Kelly, J. Miller, S. Ruff, J.L. Lehr, and A. N. Belo et al. 2006. The globally competent engineer: Working effectively with people who define problems differently. *Journal of Engineering Education* 95 (2): 107–122.

Doyle, Jack. 2002. *Riding the Dragon: Royal Dutch Shell and the Fossil Fire.* Washington, DC: Environmental Health Fund.

Duffy, J., E. Tsang, and S. Lord. 2000. Service-learning in engineering: What, why, and how? Paper presented at the American Society for Engineering Education Annual Meeting, St. Louis, MO.

Duncan, D. E. 2006. The pollution within. *National Geographic,* October. http://www7.nationalgeographic.com/ngm/0610/feature4/.

Eaton, D.L., R. B Daroff, H. Autrup, J. Bridges, L. G. Costa, J. Coyle, G. McKhann, W. C. Mobley, L. Nadel, D. Neubert, R. Schulte-Hermann, and P. S. Spencer. 2008. Review of the toxicology of chlorpyrifos with an emphasis on human exposure and neurodevelopment. *Critical Reviews in Toxicology* 38 (2): 1–125.

Edwards, Paul. 1992. Global climate science, uncertainty and politics: Data-laden models, model-filtered data. *Science as Culture* 8:437–472.

Edwards, Paul. 2010. *A Vast Machine: Computer Models, Climate Data. and the Politics of Global Warming.* Cambridge, MA: MIT Press.

Egan, Michael. 2007. *Barry Commoner and the Science of Survival: The Remaking of American Environmentalism.* Cambridge, MA: MIT Press.

Energy Forum. 2002. *Milestones Report, Issue 5.* Colombo: Energy Forum.

Environmental Defence. 2005. *Toxic Nation: A Report on Pollution in Canadians.* http://www.environmentaldefence.ca/toxicnation/resources/publications.htm. Toronto: Environmental Defence, Canada.

Environmental Mutagen Society. 2005. Marvin S. Legator. http://www.ems-us.org/who_we_are/memorial/mlegator.asp.

Environmental Working Group. 2003a. Body burden: The pollution in people. http://archive.ewg.org/reports/bodyburden1/. Oakland, CA, and Washington, DC: Environmental Working Group.

Environmental Working Group. 2003b. Mother's milk. http://www.ewg.org/reports/mothersmilk. Oakland, CA, and Washington, DC: Environmental Working Group.

Environmental Working Group. 2005. Body burden—the pollution in newborns. Oakland, CA, and Washington, DC: Environmental Working Group. http://archive.ewg.org/reports/bodyburden2/execsumm.php.

Epstein, Steven. 1996. *Impure Science: AIDS, Activism, and the Politics of Knowledge.* Berkeley: University of California Press.

Esposito, John C. 1970. *Vanishing Air: The Ralph Nader Study Group Report on Air Pollution.* New York: Grossman.

Faber, Daniel. 2008. *Capitalizing on Environmental Injustice: The Polluter-Industrial Complex in the Age of Globalization.* New York: Rowman and Littlefield.

Fiorino, Daniel J. 1990. Citizen participation and environmental risk: A survey of institutional mechanisms. *Science, Technology & Human Values* 15:226–243.

Fischer, D. 2005. What's in you? *Oakland Tribune*, March 10.www.insidebayarea.com/bodyburden.

Fischer, Frank. 1990. *Technocracy and the Politics of Expertise.* Newbury Park, CA: Sage.

Fischer, Frank. 1999. Technological deliberation in a democratic society: The case for participatory inquiry. *Science & Public Policy* 26:294–302.

Fischer, Frank. 2000. *Citizens, Experts, and the Environment: The Politics of Local Knowledge.* Durham, NC: Duke University Press.

Flyvbjerg, Bent. 1998. *Rationality and Power: Democracy in Practice.* Chicago: University of Chicago Press.

Fortun, Kim. 2001. *Advocacy after Bhopal: Environmentalism, Disaster, New Global Orders*. Chicago: University of Chicago Press.

Fortun, Kim. 2005. From Bhopal to the informating of environmental health: Risk communication in historical perspective. In Gregg Mitman, Michelle Murphy, and Christopher Sellers, eds., *Landscapes of Exposure: Knowledge and Illness in Modern Environments. Osiris*, vol. 19, 283–296. Chicago: University of Chicago Press.

Fortun, Kim. 2009. Environmental right-to-know and the transmutations of law. In Austin Sarat, ed., *Catastrophe: Law, Politics, and the Humanitarian Impulse*. Amherst: University of Massachusetts Press.

Fortun, Kim, and Michael Fortun. 2005. Scientific imaginaries and ethical plateaus in contemporary U.S. toxicology. *American Anthropologist* 107 (1): 43–54.

Fortun, Mike. 2005. For an ethics of promising, or, a few kind words about James Watson. *New Genetics & Society* 24 (2): 157–173.

Foster, Andrea, Peter Fairley, and Rick Mullin. 1998. Scorecard hits home: Web site confirms Internet's reach. *Chemical Week* 160 (21): (June 3): 24– 26.

Foucault, M., and C. Gordon. 1980. *Power/Knowledge: Selected Interviews and Other Writings, 1972–1977*. New York: Pantheon Books.

Frankena, Frederick. 1992. *Strategies of Expertise in Technical Controversies: A Study of Wood Energy Development*. Bethlehem, PA: Lehigh University Press.

Frickel, Scott. 2004. Just science? Organizing scientist activism in the US environmental justice movement. *Science as Culture* 13 (4): 449–469.

Frickel, Scott. 2005. *Chemical Consequences: Environmental Mutagens, Scientist Activism, and the Rise of Genetic Toxicology*. New Brunswick, NJ: Rutgers.

Frickel, Scott. 2008. On missing New Orleans: Lost knowledge and knowledge gaps in an urban hazardscape. *Environmental History* 13:643–650.

Frickel, Scott. 2010. Shadow mobilization for environmental and health justice. In Jane Banaszak-Holl, Sandra R. Levitsky, and Mayer Zald, eds., *Social Movements and the Transformation of American Health Care*, 171–187. London: Oxford University Press.

Frickel, Scott, Sahra Gibbon, Jeff Howard, Joanna Kempner, Gwen Ottinger, and David Hess. 2010. Undone science: Charting social movement and civil society challenges to research agenda setting. *Science, Technology & Human Values* 35 (4):444–473.

Frickel, Scott, and Kelly Moore, eds. 2006. *The New Political Sociology of Science: Institutions, Networks, and Power*. Madison: University of Wisconsin Press.

Frugal, C., S. Kalhok, E. Loring, and S. Smith. 2003. *Knowledge in Action: Canadian Arctic Contaminants Assessment Report II*. www.ainc-inac.gc.da. Ottawa, Canada: Minister of Indian Affairs and Northern Development.

Fung, A., and D. O'Rourke. 2000. Reinventing environmental regulation from the grassroots up: Explaining and expanding the success of the toxics release inventory. *Environmental Management* 25 (2): 115–127.

Galusky, Wyatt. 2000. The promise of conservation biology. *Organization & Environment* 13:226–232.

Galusky, Wyatt. 2004. *Virtually Uninhabitable: A Critical Analysis of Digital Environmental Anti-toxics Activism.* Doctoral dissertation, Virginia Polytechnic Institute and State University.

Gates, B. 1999. *Business @ the Speed of Thought: Using a Digital Nervous System.* New York: Warner Books.

Gauna, Eileen. 1998. The environmental justice misfit: Public participation and the paradigm paradox. *Stanford Environmental Law Journal* 3:4–72.

Gelbspan, Ross. 2005. *Boiling Point: How Politicians, Big Oil and Coal, Journalists, and Activists Have Fueled a Climate Crisis—And What We Can Do to Avert Disaster.* New York: Basic Books.

Gibbs, Lois. 1982. *Love Canal: My Story.* Albany: SUNY Press.

Gieryn, Thomas F. 1983. Boundary-work and the demarcation of science from non-science: Strains and interests in professional ideologies of scientists. *American Sociological Review* 48 (6): 781–795.

Gieryn, Thomas F. 1999. *Cultural Boundaries of Science: Credibility on the Line.* Chicago: University of Chicago Press.

Gilmore, Thomas, Jim Krantz, and Rafael Ramirez. 1986. Action based modes of inquiry and the host-researcher relationship. *Consultation* 5 (3): 160–176.

Glenna, Leland L., William B. Lacy, Rick Welsh, and Dina Biscotti 2007. University administrators, agricultural biotechnology, and academic capitalism: Defining the public good to promote university-industry relationships. *Sociological Quarterly* 48 (1): 141–163.

Goff, R. M., C. Williams, J. P. Terpenny, K. Gilbert, T. Knott, and J. Lo. 2010. ROXIE: Real Outreach Experiences in Engineering: First-year engineering students designing for community partners. *International Journal of Engineering Education* 26 (2): 349–358.

Goffman, Erving. 1959. *The Presentation of Self in Everyday Life.* New York: Doubleday.

Goldstein, B. D. 2005. Advances in risk assessment and communication. *Annual Review of Public Health* 26:141–163.

Gordon, Holly D., and Keith I. Harley. 2005. Environmental justice and the legal system. In David N. Pellow and Robert J. Brulle, eds., *Power, Justice, and the Environment: A Critical Appraisal of the Environmental Justice Movement,* 153–170. Cambridge, MA: MIT Press.

Gottlieb, Robert. 1993. *Forcing the Spring: The Transformation of the American Environmental Movement.* Washington, DC: Island Press.

Gould, Kenneth A. 1994. Legitimacy and growth in the balance: The role of the state in environmental remediation. *Industrial & Environmental Crisis Quarterly* 8 (3): 237–256.

Graham, Mary. 2000. Regulation by shaming: Sometimes the best way to get companies to change is to make them come clean. *Atlantic Monthly* 285:36–41.

Greenpeace International. 2005. A present for life–hazardous chemicals in umbilical cord blood. Amsterdam: Greenpeace International. http://www.greenpeace .org/eu-unit/press-centre/reports/a-present-for-life.

Guston, David. 1999. Evaluating the First U.S. Consensus Conference: The Impact of the "Citizens" Panel on Telecommunications and the Future of Democracy. *Science, Technology & Human Values* 24 (4):451–482.

Hadden, Susan. 1989. *A Citizen's Right to Know: Risk Communication and Public Policy*. Boulder, CO: Westview Press.

Haraway, Donna. 1988. Situated knowledges: The science question in feminism and the privilege of partial perspective. *Feminist Studies* 14, no. 3 (Autumn): 575–599.

Harding, Sandra. 1991. *Whose Science? Whose Knowledge?* Ithaca, NY: Cornell University Press.

Harding, Sandra. 1998. *Is Science Multicultural? Postcolonialisms, Feminisms, and Epistemologies*. Bloomington: Indiana University Press.

Harner, John, Kee Warner, John Pierce, and Tom Huber. 2002. Urban environmental justice indices. *Professional Geographer* 54 (3): 318–331.

Harnly, Martha, Robert McLaughlin, Asa Bradman, Meredith Anderson, and Robert Gunier 2005. Correlating agricultural use of organophosphates with outdoor air concentrations: A particular concern for children. *Environmental Health Perspectives* 113 (9): 1184–1189.

Harris, Stuart, and Barbara Harper. 1997. A Native American exposure scenario. *Risk Analysis* 17:789–795.

Harris, Stuart, and Barbara Harper. 2000. Using eco-cultural dependency webs in risk assessment and characterization of risks to tribal health and cultures. *Environmental Science and Pollution Research* 2:91–100.

Harrison, Jill Lindsey. 2006. "Accidents" and invisibilities: Scaled discourses and the naturalization of regulatory neglect in California's pesticide drift conflict. *Political Geography* 25:506–529.

Harrison, Jill Lindsey. 2008. Abandoned bodies and spaces of sacrifice: Pesticide drift activism and the contestation of neoliberal environmental politics in California. *Geoforum* 39:1197–1214.

Harvey, David. 2006. *Spaces of Global Capitalism: Towards a Theory of Uneven Geographical Development*. New York: Verso.

Head, Rebecca A. 1995. Health-based standards: What role in environmental justice? In Bunyan Bryant, ed., *Environmental Justice: Issues, Policies, Solutions*, 45–56. Washington, DC: Island Press.

Henke, Christopher R. 2006. Changing ecologies: Science and environmental politics in agriculture. In S. Frickel and K. Moore, eds., *The New Political Sociology of Science: Institutions, Networks, and Power*, 215–243. Madison: University of Wisconsin Press.

Hess, David J. 2007. *Alternative Pathways in Science and Industry: Activism, Innovation, and the Environment in an Era of Globalization.* Cambridge, MA: MIT Press.

Hess, David J. 2009. *Localist Movements in a Global Economy: Sustainability, Justice, and Urban Development in the United States.* Cambridge, MA: MIT Press.

Hilgartner, Stephen. 2000. *Science on Stage: Expert Advice as Public Drama.* Stanford, CA: Stanford University Press.

Hoffmann-Riem, Holger, and Brian Wynne. 2002. In risk assessment, one has to admit ignorance. *Nature* 416 (6877): 123.

Holland, Dorothy, and Margaret Eisenhart. 1991. *Educated in Romance: Women, Achievement, and College Culture.* Chicago: University of Chicago Press.

Holland, Dorothy, and Jean Lave. 2001. *History in Person: Enduring Struggles, Contentious Practice, Intimate Identities.* Santa Fe, NM: School of American Research Press.

Hooper, Kin, and Jianwen She. 2003. Lessons from the polybrominated diphenyl ethers (PBDEs): Precautionary principle, primary prevention, and the value of community-based body-burden monitoring using breast milk. *Environmental Health Perspectives* 111:109–114.

Howard, Jeff. 2004. *Toward Intelligent, Democratic Steering of Chemical Technologies: Evaluating Industrial Chlorine Chemistry as Environmental Trial and Error.* Doctoral dissertation, Rensselaer Polytechnic Institute.

Huber, Peter. 1998. Saving the environment from the environmentalists. *Commentary.* Manhattan Institute of Policy Research 105 (5) April.

Iles, Alastair. 2007. Identifying environmental health risks in consumer products: Non-governmental organizations and civic epistemologies. *Public Understanding of Science* 16:371–391.

Irwin, Alan. 1995. *Citizen Science: A Study of People, Expertise, and Sustainable Development.* London: Routledge.

Irwin, Alan. 2004. Rather than representing citizens as risk-averse, we should be engaging more with what people want from technical change. *Spiked Online Debate.* http://www.spiked-online.com/articles/0000000CA375.htm.

Irwin, Alan, Alison Dale, and Denis Smith. 1996. Science and Hell's Kitchen: The local understanding of hazard issues. In A. Irwin and B. Wynne, eds., *Misunderstanding Science? The Public Reconstruction of Science and Technology,* 47–64. Cambridge: Cambridge University Press.

Irwin, Alan, and Brian Wynne. 1996. *Misunderstanding Science? The Public Reconstruction of Science and Technology.* Cambridge: Cambridge University Press.

Israel, B., and A. J. Schultz, E. A. Parker, A. B. Becker, A. J. Allen, and J. R. Guzman. 2003. Critical issues in developing and following community-based participatory research principles. In M. Minkler and N. Wallerstein, eds.,

Community-Based Participatory Research for Health, 53–76. San Francisco: Jossey-Bass.

Jacobson, Joseph L., and Sandra W. Jacobson. 1996. Intellectual impairment in children exposed to polychlorinated biphenyls in utero. *New England Journal of Medicine* 335 (11): 783–789.

Jasanoff, S. 1987. Contested boundaries in policy-relevant science. *Social Studies of Science* 17:195–230.

Jasanoff, S. 1990. *The Fifth Branch: Science Advisers as Policymakers.* Cambridge, MA: Harvard University Press.

Jasanoff, Sheila. 1992. Science, politics, and the renegotiation of expertise at EPA. *Osiris* 7:1–23.

Jasanoff, Sheila. 1995. Procedural choices in regulatory science. *Technology in Society* 17 (3): 279–293.

Jepperson, Ronald L. 1991. Institutions, institutional effects, and institutionalism. In W. W. Powell and P. J. DiMaggio, eds., *The New Institutionalism in Organizational Analysis*, 143–163. Chicago: University of Chicago Press.

Kane, Eugene. 2005. Racial struggles of past still present today. *Milwaukee Journal Sentinel*, April 24, 2005.

Kaplan, Sheila. 2010. Reform of chemical law collapses as industry flexes its muscles. *Politics Daily*, October 13, 2010. http://www.politicsdaily.com/2010/10/13/reform-of-toxic-chemicals-law-collapses-as-industry-flexes-its-m/.

Keck, Margaret E., and Kathryn Sikkink. 1998. *Activists beyond Borders: Advocacy Networks in International Politics.* Ithaca, NY: Cornell University Press.

Keller, Evelyn Fox. 2002. *Making Sense of Life: Explaining Biological Development with Models, Metaphors, and Machines.* Cambridge, MA: Harvard University Press.

Kezar, Adrianna, and Jaime Lester. 2009. Promoting grassroots change in higher education–the promise of virtual networks. *Change* 41 (2): 44–51.

Kishi, Misa. 2005. The health impacts of pesticides: What do we now know? In J. Pretty, ed., *The Pesticide Detox: Towards a More Sustainable Agriculture*, 23–38. Sterling, VA: Earthscan.

Kleinman, Daniel Lee, and Steven P. Vallas. 2006. Contradiction and convergence: Universities and industry in the biotechnology field. In S. Frickel and K. Moore, eds., *The New Political Sociology of Science: Institutions, Networks, and Power*, 35–62. Madison: University of Wisconsin Press.

Knorr, Karin D. 1981. The Manufacture of Knowledge: An Essay on the Constructivist and Contextual Nature of Science. Oxford: Pergamon Press.

Kohler, Robert E. 2002. *Landscapes and Labscapes: Exploring the Lab-Field Border in Biology.* Chicago: University of Chicago Press.

Krupp, Fred. 1999. *A Letter from EDF's Executive Director.* http://www.scorecard.org/about/about-why.tcl.

Kubasek, Nancy, and Gary Silverman. 1999. *Environmental Law*. Upper Saddle River, NJ: Prentice Hall.

Kuehn, Robert. 1996. The environmental justice implications of quantitative risk assessment. *University of Illinois Law Review* 103:1–67.

Kuhn, Thomas. [1962] 1996. *The Structure of Scientific Revolutions*. 3rd ed. Chicago: University of Chicago Press.

Laraña, Enrique, Hank Johnston, and Joseph R. Gusfield, eds. 1994. *New Social Movements: From Ideology to Identity*. Philadelphia: Temple University Press.

Latour, Bruno, and Steve Woolgar. [1979] 1986. *Laboratory Life: The Construction of Scientific Facts*. 2nd ed. Princeton, NJ: Princeton University Press.

Lave, Jean, and Etienne Wenger. 1991. *Situated Learning: Legitimate Peripheral Participation*. Cambridge: Cambridge University Press.

Lee, Sharon, Robert McLaughlin, Martha Harnly, Robert Gunier, and Richard Kreutzer 2002. Community exposures to airborne agricultural pesticides in California: Ranking of inhalation risks. *Environmental Health Perspectives* 110 (12): 1175–1184.

Lehr, J. L., E. McCallie, S. R. Davies, B. R. Caron, B. Gammon, and S. Duensing. 2007. The value of "dialogue events" as sites of learning: An exploration of research and evaluation frameworks. *International Journal of Science Education* 29 (12): 1467–1487.

Leitner, Helga, Jamie Peck, and Eric S. Sheppard. 2007. *Contesting Neoliberalism: Urban Frontiers*. New York: Guilford Press.

Lerner, Steve. 2005. *Diamond: A Struggle for Environmental Justice in Louisiana's Chemical Corridor*. Cambridge, MA: MIT Press.

Levine, Adeline. 1982. *Love Canal: Science, Politics, and People*. Lexington, MA: Lexington Books.

Lewis, Sanford, Brian Keating, and Dick Russell. 1992. *Inconclusive by Design: Waste, Fraud and Abuse in Federal Environmental Health Research*. Environmental Health Network.

Lioy, Paul. J., Natalie C. Freeman, and James R Millette. 2002. Dust: A metric for use in residential and building exposure assessment and source characterization. *Environmental Health Perspectives* 110 (10): 969–983.

London, Jonathan K., Julie Sze, and Raoul S. Liévanos. 2008. Problems, promise, progress, and perils: Critical reflections on environmental justice policy implementation in California. *UCLA Journal of Environmental Law & Policy* 26 (2): 255–289.

Lopez, Russ. 2002. Segregation and Black/White differences in exposure to air toxics in 1990. *Environmental Health Perspectives (Supplements)* 110 (2): 289–295.

Luke, Timothy. 1995. On environmentality: Geo-power and geo-knowledge in the discourses of contemporary environmentalism. *Cultural Critique* 31: 57–81.

Lynch, Michael. 1985. *Art and Artifact in Laboratory Science: A Study of Shop Work and Shop Talk in a Research Laboratory*. London: Routledge and Kegan Paul.

MacKenzie, Donald, and Judy Wajcman. 1999. *The Social Shaping of Technology*. 2nd ed. London: McGraw Hill Education.

Marshall, C. 2005. New emission rule for Bay Area refineries. *New York Times*, July 21, 21.

Martin, Brian. 1996. *Confronting the Experts*. Albany: State University of New York Press.

Martin, Brian. 2006. Strategies for alternative science. In Scott Frickel and Kelly Moore, eds., *The New Political Sociology of Science: Institutions, Networks, and Power*, 272–298. Madison: University of Wisconsin Press.

Martin, Philip L., and J. Edward Taylor. 1998. Poverty amid prosperity: Farm employment, immigration, and poverty in California. *American Journal of Agricultural Economics* 80 (5): 1008–1014.

Marwell, Nicole P. 2004. Privatizing the welfare state: Nonprofit community-based organizations as political actors. *American Sociological Review* 69 (2): 265–291.

Matteson, Patricia, Larry Wihoit, and Mark Robertson. 2007. Environmental justice pilot project pest management assessment: Soil fumigant and organophosphate insecticide use and alternatives—Parlier, Fresno County, California. California Department of Pesticide Regulation. http://www.cdpr.ca.gov/docs/pestmgt/pubs/ej_pma_final10.pdf.

McCormick, Sabrina, Phil Brown, and Stephen Zavestoski. 2003. The personal is scientific, the scientific is political: The public paradigm of the environmental breast cancer movement. *Sociological Forum* 18 (4): 545–576.

McGraw, Joseph, and Donald P. Waller. 2008. Fish ingestion and congener specific polychlorinated biphenyl and p,p'-dichlorodiphenyldichloroethylene serum concentrations in a Great Lakes cohort of pregnant African American women. *Environment International* 35 (3): 557–565.

McLaughlan, R. G. 2007. Instructional strategies to educate for sustainability in technology assessment. *International Journal of Engineering Education* 23 (2): 201–208.

Metcalf, S. W., and K. G. Orloff. 2004. Biomarkers of exposure in community settings. *Journal of Toxicology and Environmental Health*, part A, 67 (8–10): 715–726.

Michaels, David. 2008. *Doubt Is Their Product: How Industry's Assault on Science Threatens Your Health*. New York: Oxford University Press.

Mihelcic, J. R., L. D. Phillips, and D. W. Watkins. 2006. Integrating a global perspective into education and research: Engineering international sustainable development. *Environmental Engineering Science* 23 (3): 426–438.

Miller, Jeffrey. 1987. *Citizen Suits: Private Enforcement of Federal Pollution Control Laws*. Washington, DC: Environmental Law Institute.

Mohai, Paul, and Bunyan Bryant. 1992. Environmental racism: Revising the evidence. In P. M. B. Bryant, ed., *Race and the Incidence of Environmental Hazards: A Time for Discourse*. Boulder, CO: Westview.

Moore, Colleen F. 2003. *Silent Scourge: Children, Pollution and Why Scientists Disagree*. New York: Oxford University Press.

Moore, Kelly. 2008. *Disrupting Science: Social Movements, American Scientists, and the Politics of the Military, 1945–1975*. Princeton, NJ: Princeton University Press.

Moore, Kelly, Scott Frickel, David Hess, and Daniel L. Kleinman. Forthcoming. Science and neoliberal globalization: A political sociological approach. *Theory and Society*.

Morello-Frosch, Rachel, Julia Brody, Phil Brown, Rebecca Gasior Altman, Ruthann Rudel, and Carla Perez. 2009. Toxic ignorance and right-to-know in biomonitoring results communication: A survey of scientists and study participants. *Environmental Health* 8 (February 28): 6.

Morello-Frosch, R., M. Pastor, J. Sadd, C. Porras, and M. Prichard. 2005. Citizens, science, and data judo: Leveraging community-based participatory research to build a regional collaborative for environmental justice in Southern California. In E. E. Barbara Israel, Amy Shultz, and Edith Parker, eds., *Methods for Conducting Community-Based Participatory Research in Public Health*. San Francisco: Jossey-Bass.

Morello-Frosch, Rachel, Stephen Zavestoski, Phil Brown, Rebecca Gasior Altman, Sabrina McCormick, and Brain Mayer.. 2006. Embodied Health Movements: Responses to a 'scientized' world . In Kelly Moore and Scott Frickel, eds., *The New Political Sociology of Science: Institutions, Networks, and Power*. Madison: University of Wisconsin Press, 244–271

Mosse, David. 2001. "People's knowledge," participation and patronage: Operations and representations in rural development. In Bill Cooke and Uma Kothari, eds., *Participation: The New Tyranny?*, 16–35. London: Zed Books.

Nadasdy, Paul. 1999. The politics of TEK: Power and the "integration" of knowledge. *Arctic Anthropology* 36:1–18.

National Academy of Engineering. 2004. *The Engineer of 2020: Visions of Engineering in the New Century*. Washington, DC: National Academies Press.

National Academy of Science. 2008. *Public Participation in Environmental Assessment and Decision Making*. Washington, DC: National Academies Press.

National Institutes of Health. 1979. *The Belmont Report: Ethical Principles and Guidelines for the Protection of Human Subjects of Research*. Bethesda, MD: National Institutes of Health, Office of Human Subjects Research. http://ohsr .od.nih.gov/guidelines/belmont.html.

National Research Council. 1983. *Risk Assessment in the Federal Government: Managing the Process*. Washington, DC: The National Academies Press.

National Research Council. 2000. *Toxicological Effects of Methylmercury*. Washington, DC: The National Academies Press.

National Research Council. 2009. *Science and Decisions: Advancing Risk Assessment*. Washington, DC: The National Academies Press.

Nieusma, Dean. 2004a. Alternative design scholarship: Working toward appropriate design. *Design Issues* 20:13–24.

Nieusma, Dean. 2004b. *The Energy Forum of Sri Lanka: Working toward Appropriate Expertise*. Doctoral dissertation, Rensselaer Polytechnic Institute.

Nieusma, Dean. 2007. Challenging knowledge hierarchies: Working toward sustainable development in Sri Lanka's energy sector. *Sustainability: Science, Practice, & Policy* 3 (1) (Spring): 32–44.

Nieusma, Dean, and Donna Riley. 2010. Designs on development: Engineering, globalization, and social justice. Engineering Studies 2, no. 1 (April): 29–59.

Norén, K., and D. Meironyté. 2000. Certain organochlorine and organobromine contaminates in Swedish human milk in perspective of past 20–30 years. *Chemosphere* 40 (9-11):1111–1123.

Norris, Pippa. 2001. *Digital Divide: Civic Engagement, Information Poverty, and the Internet Worldwide*. New York: Cambridge University Press.

Novak, Bill. 2006. Catch of the day: Good info–Fish advisories need to get out in three languages. *Capital Times*, September 15.

Nowotny, Helga, Peter Scott, and Michael Gibbons. 2001. *Re-thinking Science: Knowledge and the Public in an Age of Uncertainty*. Cambridge: Polity Press.

NTCF (National Toxics Campaign Fund). 1993. *National Toxics Campaign: Some Reflections, Thoughts for the Movement*. Photocopy.

Nugent, Angela. 1985. Organizing trade unions to combat disease: The Workers' Health Bureau, 1921–1928. *Labor History* 26 (Summer): 423–446.

Obituary for Marvin S. Legator. 2005. *Houston Chronicle*, July 21, B7.

O'Brien, Rory. 2001. An overview of the methodological approach of action research. In R. Richardson, ed., *Theory and Practice of Action Research*. João Pessoa, Brazil: Universidade Federal da Paraíba.

O'Neill, Christine. 2000. Variable justice: Environmental standards, contaminated fish, and "acceptable" risk to Native peoples. *Stanford Environmental Law Journal* 19:3–120.

Ong, Elisa, and Stanton Glantz. 2001. Constructing "sound science" and "good epidemiology": Tobacco, lawyers, and public relations firms. *American Journal of Public Health* 91 (11):1749–1757.

Open Letter. 1990. *Not Man Apart: The Newsmagazine of Friends of the Earth*, USA, 20 (2): 15–16.

O'Rourke, Dara, and Gregg P. Macey. 2003. Community environmental policing: Assessing new strategies of public participation in environmental regulation. *Journal of Policy Analysis and Management* 22:383–414.

Ottinger, Gwen. 2010. Buckets of resistance: Standards and the effectiveness of citizen science. *Science, Technology & Human Values* 35 (2): 244–270.

Oudshoorn, Nelly, and Trevor Pinch, eds. 2003. *How Users Matter: The Co-construction of Users and Technology.* Cambridge, MA: MIT Press.

Overdevest, C., and B. Mayer. 2008. Harnessing the power of information through community monitoring: Insights from social science. *Texas Law Review* 86 (7): 1493–1526.

Pardo, Mary S. 1998. *Mexican American Women Activists: Identity and Resistance in Two Los Angeles Communities.* Philadelphia: Temple University Press.

Pellow, David N. 2001. Environmental justice and the political process: Movements, corporations, and the state. *Sociological Quarterly* 42 (1): 47–67.

Pellow, David N. 2007. *Resisting Global Toxics: Transnational Movements for Environmental Justice.* Cambridge, MA: MIT Press.

Pellow, David N., and Robert J. Brulle, eds. 2005. *Power, Justice, and the Environment: A Critical Appraisal of the Environmental Justice Movement.* Cambridge, MA: MIT Press.

Peterson, T. H. 2009. Engaged scholarship: Reflections and research on the pedagogy of social change. *Teaching in Higher Education* 14 (5): 541–552.

Petersen, Dana, Meredith Minkler, Victoria Breckwich Vasquez, and Andrea Corage Baden. 2006. Community based participatory research as a tool for policy change: A case study of Southern California Environmental Justice Collaborative. *Review of Policy Research* 23 (2):339–354.

Ngo, Mai A., Kent Pinkerton, Sandra Freeland, Michael Geller, Walter Ham, Steven Cliff, and Laurie E. Hopkins. 2010. Airborne particles in San Joaquin Valley may affect human health. *California Agriculture,* January–March: 12–16.

Povinelli, Elizabeth. 1997. Reading ruptures, rupturing readings: Mabo and the cultural politics of activism. *Social Analysis* 41:20–28.

Povinelli, Elizabeth. 1998. The state of shame: Australian multiculturalism and the crisis of indigenous citizenship. *Critical Inquiry* 24:575–610.

Povinelli, Elizabeth. 2002. *The Cunning of Recognition: Indigenous Australian Multiculturalism.* Durham, NC: Duke University Press.

Powell, Jim, and Maria Powell. 2008. *The State of Shoreline Fishing in Dane County: A Report on Fishing, Fish Consumption, and Public Health Advisories.* Madison, WI: Madison Environmental Justice Organization.

Powell, Maria C. 2004. *To Know or Not to Know? Perceived Uncertainty and Information Seeking and Processing about Contaminated Great Lakes Fish.* Doctoral dissertation, University of Wisconsin–Madison.

Powell, Maria, Sharon Dunwoody, Robert Griffin, and Kurt Neuwirth. 2007. Exploring lay uncertainty about an environmental health risk. *Public Understanding of Science* (Bristol, England) 17 (3): 329–348.

Powell, Maria, Ly V. Xiong, and James Powell. 2009. *Fish Consumption Advisory Sign Survey Analysis: A Report on Angler Awareness of Fish Advisories in Dane County.* Madison, WI: Madison Environmental Justice Organization.

Public Health Madison & Dane County. 2008. *Fish Consumption Advisory Education Efforts in Dane County.* Madison, WI: Public Health Madison and Dane County, July 23.

Pulido, Laura. 1996. *Environmentalism and Economic Justice: Two Chicano Struggles in the Southwest.* Tucson: University of Arizona Press.

Quandt, Sara A., and Alicia M. Doran, Pamela Rao, Jane A. Hoppin, Beverely M. Snively, and Thomas A. Arcury. 2004. Reporting pesticide assessment results to farmworker families: Development, implementation, and evaluation of a risk communication strategy. *Environmental Health Perspectives* 112 (5): 636–642.

Ranco, Darren J. 2000. *Environmental Risk and Politics in Eastern Maine: The Penobscot Nation and the Environmental Protection Agency.* Doctoral dissertation, Harvard University.

Rawls, John. 2001. *Justice as Fairness: A Restatement.* Ed. Erin Kelly. Cambridge, MA: Belknap Press of Harvard University Press.

Riley, D., and A. H. Bloomgarden. 2006. Learning and service in engineering and global development. *International Journal for Service Learning in Engineering* 2 (1): 48–59.

Rose, Carol. 1986. The comedy of the commons: Custom, commerce, and inherently public property. *University of Chicago Law Review* 53 (3): 711–778.

Rose, Nikolas. 1996. *Inventing Our Selves: Psychology, Power, and Personhood.* Cambridge: Cambridge University Press.

Rosenbaum, Walter A. 2008. *Environmental Politics and Policy.* Washington, DC: CQ Press.

Rosner, David, and Gerald Markowitz. 1987. Safety and health as a class issue: The Workers' Health Bureau of America during the 1920s. In David Rosner and Gerald Markowitz, eds., *Dying for Work: Workers' Safety and Health in Twentieth Century America,* 53–64. Bloomington: Indiana University Press.

Rowe, G., and L. J. Frewer. 2005. A typology of public engagement mechanisms. *Science, Technology & Human Values* 30 (2): 251–290.

Rudel, Ruthann A., David E. Camann, John D. Spengler, Leo R. Korn, and Julia G. Brody 2003. Phthalates, alkylphenols, pesticides, polybrominated diphenyl ethers, and other endocrine-disrupting compounds in indoor air and dust. *Environmental Science & Technology* 37 (20): 4543–4553.

Schafer, Kristin S., Margaret Reeves, Skip Spitzer, and Susan E. Kegley. 2004. *Chemical Trespass: Pesticides in Our Bodies and Corporate Accountability.* San Francisco: Pesticide Action Network of North America.

Schantz, Susan, Donna M. Gasior, Elena Polverejan, Robert J. McCaffrey, Anne M. Sweeney, Harold E. B. Humphrey, and Joseph C. Gardiner. 2001. Impairments of memory and learning in older adults exposed to polychlorinated biphenyls via consumption of Great Lakes fish. *Environmental Health Perspectives* 109 (6): 605–611.

Schell, L. M., and A. M. Tarbell. 1998. A partnership study of PCBs and the health of Mohawk youth: Lessons from our past and guidelines for our future. *Environmental Health Perspectives* 106 (Suppl. 3): 833–840.

Schierow, Linda-Jo. 2009. Toxic Substances Control Act (TSCA): Implementation and new challenges. *Congressional Research Service*, CRS RL-34118, July 29: 1–39..

Schlosberg, David. 1999. *Environmental Justice and the New Pluralism: The Challenge of Difference for Environmentalism*. Oxford: Oxford University Press.

Schneider, Pat. 2008a. Health warnings at fishing holes? *Capital Times*, Madison, WI, February 12.

Schneider, Pat. 2008b. Report: Local anglers eating too much fish: Minorities more likely to exceed safety guidelines. *Capital Times*, Madison, WI, July 27.

Schrader-Frechette, Kristin. 2002. *Environmental Justice: Creating Equality, Reclaiming Democracy*. Oxford: Oxford University Press.

Schulte, P., and M. Singal. 1996. Ethical issues in the interaction with subjects and disclosure of results. In S. Coughlin and T. Beauchamp, eds., *Ethics and Epidemiology*, 178–198. New York: Oxford University Press.

Schweizer, Errol. 1999. Interview with Robert Bullard. *Earth First Journal*, July 6. http://www.ejnet.org/ej/bullard.html

Sclove, R. 1993. Technological politics as if democracy really mattered. In A. Teich, ed., *Technology and the Future*. New York: St. Martin's Press.

Sclove, Richard. 1995. *Democracy and Technology*. New York: Guilford Press.

Sclove, Richard. 2000. Town meetings on technology: Consensus conferences as democratic participation. In Daniel Lee Kleinman, ed., *Science, Technology, and Democracy*, 33–48. Albany: SUNY Press.

Sellers, Christopher C. 1997. *Hazards of the Job: From Industrial Disease to Environmental Health Science*. Chapel Hill: University of North Carolina Press.

Serres, Michel. 1983. *Hermes: Literature, Science, Philosophy*. Baltimore: Johns Hopkins University Press.

Shalowitz, D., and F. Miller. 2005. Disclosing individual results of clinical research implications of respect for participants. *Journal of the American Medical Association* 294 (6): 737–740.

Shapin, Steven. 2008. *The Scientific Life*. Chicago: University of Chicago Press.

Shepard, Peggy M., Mary E. Northridge, Swati Prakash, and Gabriel Stover. 2002. Preface: Advancing environmental justice through community based participatory action research. *Environmental Health Perspectives* 110 (S2): 139–144.

Sherman, Jennifer, Don Villarejo, Anna Gracia, Stevan McCurdy, Ketty Mobed, David Runsten, Kathy Saiki et al. 1997. *Finding Invisible Farmworkers: The Parlier Survey*. Davis: California Institute for Rural Studies.

Shuman, L. J., M. Besterfield-Sacre, and J. McGourty. 2005. The ABET "professional skills"–Can they be taught? Can they be assessed? *Journal of Engineering Education* 94 (1): 41–55.

Sightline Institute. 2004. *Flame Retardants in the Bodies of Pacific Northwest Residents: A Study of Toxic Body Burdens*. Seattle: Sightline Institute. http://www.sightline.org/.

Singleton, Chantele R., and Marvin S. Legator. 1997. Symptom survey: Initial, critical step in a comprehensive community health study plan. *Archives of Environmental Health* 52:255–256.

Sismondo, Sergio. 2004. *An Introduction to Science and Technology Studies*. Oxford: Blackwell.

Sismondo, Sergio. 2008. Science and technology studies and an engaged program. In Hackett, Edward J., Olga Amsterdamska, Michael Lynch, and Judy Wajcman, eds., *The Handbook of Science and Technology Studies, 3rd ed.*, 13–31. Cambridge, MA: MIT Press.

Sklar, Kathryn Kish. 1995. Two political cultures in the Progressive Era: The National Consumers' League and the American Association for Labor Legislation. In Linda K. Kerber, Alice Kessler-Harris, and Kathryn Kish Sklar, eds., *U.S. History as Women's History: New Feminist Essays*, 36–62. Chapel Hill: University of North Carolina Press.

Slater, C. S., R. P. Hesketh, D. Fichana, J. Henry, A. M. Flynn, and M. Abraham. 2007. Expanding the frontiers for chemical engineers in green engineering education. *International Journal of Engineering Education* 23 (2): 309–324.

Smithson, Michael. 1989. *Ignorance and Uncertainty: Emerging Paradigms*. New York: Springer-Verlag.

Star, Susan L., and James R. Griesemer. 1989. Institutional ecology, "translations," and boundary objects: Amateurs and professionals in Berkeley's Museum of Vertebrate Zoology, 1907–39. *Social Studies of Science* 19:387–420.

State of Black Madison Coalition. 2008. *The State of Black Madison Report 2008: Before the Tipping Point*. Madison, WI: Urban League of Greater Madison.

Steenport, Diane M., Henry A. Anderson, Lawrence P. Hanrahan, Claire Falk, Laurie A. Draheim, Marty S. Kanarek, and Henry Nehls-Lowe. 2000. Fish consumption habits and advisory awareness among Fox River anglers. *Wisconsin Medical Journal* 99 (8): 43–46.

Steingraber, Sandra. 1997. *Living Downstream: A Scientist's personal investigation of cancerand the Environment*. New York: Vintage.

Stocking, S. Holly. 1998. On drawing attention to ignorance. *Science Communication* 20 (1): 165–178.

Stradling, David. 1999. *Smokestacks and Progressives: Environmentalists, Engineers, and Air Quality in America, 1881–1951*. Baltimore: Johns Hopkins University Press.

Strandman, Teila, Jaana Koistinen, and Terttu Vartiainen. 2000. Polybrominated diphenyl ethers (PBDEs) in placenta and human milk. *Organohalogen Compounds* 47:61–64.

Sullivan, M., A. Kone, K. Senturia, N. Chrisman, S. Ciske, and J. Krieger. 2001. Researcher and researched—community perspectives: Toward bridging the gap. *Health Education & Behavior* 28 (2): 130–149.

Szadkowski, Adam, and Charles R. Myers. 2008. Acrolein oxidizes the cytosolic and mitochondrial thioredoxins in human endothelial cells. *Toxicology* 243 (1–2): 164–176.

Szasz, Andrew. 1994. *Ecopopulism: Toxic Waste and the Movement for Environmental Justice.* Minneapolis: University of Minnesota Press.

Sze, Julie. 2007. *Noxious New York: The Racial Politics of Urban Health and Environmental Justice.* Cambridge, MA: MIT Press.

Sze, Julie, and Jonathan London. 2008. Environmental justice at the crossroads. *Social Compass* 2:1331–1354.

Sze, J., and S. Prakash. 2004. Human genetics, environment, and communities of color: Ethical and social implications. *Environmental Health Perspectives* 112 (6): 740–745.

Taylor, Charles. 1994. *The Politics of Recognition.* Princeton, NJ: Princeton University Press.

Tesh, Sylvia N. 2000. *Uncertain Hazards: Environmental Activists and Scientific Proof.* Ithaca, NY: Cornell University Press.

Tesh, Sylvia N., and Bruce A. Williams. 1996. Identity politics, disinterested politics, and environmental justice. *Polity* 28 (3): 285–305.

Thorpe, Charles, and Steven Shapin. 2000. Who was Robert Oppenheimer? Charisma and complex organization. *Social Studies of Science* 30:545–590.

Tilden, John, Lawrence P. Hanrahan, Henry Anderson, Charles Palit, J. Olson, W. MacKenzie, and Great Lakes Sport Fish Consortium. 1997. Health advisories for consumers of Great Lakes sport fish: Is the message being received? *Environmental Health Perspectives* 105 (12): 1360–1365.

Toffolon-Weiss, Melissa, and Timmons Roberts. 2005. Who wins, who loses? Understanding outcomes of environmental injustice struggles. In D. N. Pellow and R. J. Brulle, eds., *Power, Justice, and the Environment,* 77–90. Cambridge, MA: MIT Press.

Traweek, Sharon. 1988. *Beamtimes and Lifetimes: The World of High Energy Physicists.* Cambridge, MA: Harvard University Press.

Traweek, Sharon. 2000. Faultlines. In Roddey Reid and Sharon Traweek, eds., *Doing Science + Culture: How Cultural and Interdisciplinary Studies Are Changing the Way We Look at Science and Medicine,* 21–48. New York: Routledge.

Tribal Science Council. 2002. *Organizational Overview.* Washington, DC: USEPA.

Tribal Science Council. 2003a. *Regaining and Protecting Tribal Culture through Science.* Boston: New England Tribal Environmental Training.

Tribal Science Council. 2003b. *Risk Assessment/Health Summary.* Albuquerque, NM: USEPA.

Tribal Science Council. 2003c. Tribal Traditional Lifeways. Pyramid Lake Paiute Tribe. Nevada May 13–15. http://www.epa.gov/osp/tribes/tribal/health.pdf

Tribal Science Council. 2004. *Health and Well Being Summary.* Tulalip Reservation. Washington, D.C.: USEPA.

Turner, Stephen. 2007. Political epistemology, experts and the aggregation of knowledge. *Spontaneous Generations* 1 (1) (December): 36–47.

Turner, Stephen. 2008. The social study of science before Kuhn. In Hackett, Edward J., Olga Amsterdamska, Michael Lynch, and Judy Wajcman. eds., *The Handbook of Science and Technology Studies*, 33–62. Cambridge, MA: MIT Press.

Tuttle, L. 2000. When you don't know an Erin Brokovich. *Christian Science Monitor* (Boston), May 22, 9.

Union of Concerned Scientists. 2008. *Interference at the EPA: Science and Politics at the U.S. Environmental Protection Agency.* Cambridge, MA: UCS Publications. April. http://www.ucsusa.org/assets/documents/scientific_integrity/interference-at-the-epa.pdf.

U.S. Environmental Protection Agency. 2000a. *Estimated Per Capita Fish Consumption Collected by the United States Department of Agriculture's Food Intake by Individuals.* Washington, DC: Office of Science and Technology.

U.S. Environmental Protection Agency. 2000b. *Guidance for Assessing Chemical Contaminant Data for Use in Fish Advisories, Volume 2: Risk Assessment and Fish Consumption Limits.* 3rd ed. Washington, DC: Office of Water, U.S. Environmental Protection Agency.

U.S. Environmental Protection Agency. 2006a. *Paper on Tribal Issues Related to Tribal Health and Well Being: Documenting What We Have.* Washington, DC: U.S. Environmental Protection Agency.

U.S. Environmental Protection Agency. 2006b. *24-Hour PM2.5 Standards—Region 5 Recommendations and EPA Responses.* http://www.epa.gov/pmdesignations/2006standards/rec/region5t.htm.

U.S. Environmental Protection Agency, Office of Pollution Prevention and Toxics. 2007. *Overview: Office of Pollution Prevention and Toxics.* http://www.epa.gov/oppt/pubs/oppt101c2.pdf.

U.S. Environmental Protection Agency. 2008. *Area Designations for 2006 24-Hour Fine Particle (PM2.5) Standards—August 19, 2008.* http://www.epa.gov/pmdesignations/2006standards/documents/2008-08/rectable.htm.

U.S. Environmental Protection Agency. 2009. *Definitions: Environmental Justice.* http://www.epa.gov/environmentaljustice/.

U.S. Government Accountability Office (GAO). 2000. *Toxic Chemicals: Long-Term Coordinated Strategy Needed to Measure Exposures in Humans.* HEHS-00-80. Washington, D.C.: U.S. Government Accountability Office.

U.S. Government Accountability Office (GAO). 2005a. *Chemical Regulation: Options Exist to Improve EPA's Ability to Assess Health Risks and Manage Its Chemical Review Program.* GAO-05-458. Washington, D.C.: U.S. Government Accountability Office.

U.S. Government Accountability Office (GAO). 2005b. *Environmental Justice: EPA Should Devote More Attention to Environmental Issues When Developing Clean Air Rules.* GAO-05-289. Washington, D.C.: U.S. Government Accountability Office.

U.S. Government Accountability Office (GAO). 2006. *Chemical Regulation: Actions Are Needed to Improve the Effectiveness of EPA's Chemical Review Program.* GAO-06-1032T. Washington, D.C.: U.S. Government Accountability Office.

U.S. Government Accountability Office (GAO). 2007. *Environmental Justice: Measurable Benchmarks Needed to Gauge EPA Progress in Correcting Past Problems.* GAO-07-11402. Washington, DC: U.S. Government Accountability Office.

U.S. Government Accountability Office (GAO). 2008. *Toxic Chemicals: EPA's New Assessment Process Will Increase Challenges EPA Faces in Evaluating and Regulating Chemicals.* GAO-08-743T. Washington, DC: U.S. Government Accountability Office.

U.S. Government Accountability Office (GAO). 2009. *Chemical Regulation: Observations on Improving the Toxic Substances Control Act.* GAO-10-292T. Washington, D.C.: U.S. Government Accountability Office.

Usher, Peter J., Maureen Baikie, Marianne Demmer, Douglas Nakashima, Marc G. Stevenson, and Mark Stiles 1995. *Communicating about Contaminants in Country Foods: The Experience of Aboriginal Communities.* Ottawa, Canada: Inuit Tapiriit Kanatami.

Villarejo, Don. 2000. *Suffering in Silence: A Report on the Health of California Agricultural Workers.* Davis: California Institute for Rural Studies.

Wagner, W. E. 1997. Choosing ignorance in the manufacture of toxic products. *Cornell Law Review* 82 (4): 773–855.

Walker, Richard A. 2006. *The Conquest of Bread: 150 Years of Agribusiness in California.* New York: New Press.

Wallerstein, N., and B. Duran. 2003. The conceptual, historical, and practice roots of community based participatory research and related participatory traditions. In M. Minkler and N. Wallerstein, eds., *Community Based Participatory Research for Health,* 27–52. San Francisco: Jossey-Bass.

Warmerdam, Mary-Ann. 2007a. Interview. Sacramento, CA. May 22.

Warmerdam, Mary-Ann. 2007b. Parlier project status report—update on projected release of final report. California Department of Pesticide Regulation. http://www.cdpr.ca.gov/docs/envjust/pilot_proj/parlier_ltr_oct07.pdf.

Warner, Jennifer. 2004. America's top asthma capitals: Worst cities for people with asthma announced. *WebMD Health News,* March 16.

Wassom, John S. 1989. Origins of genetic toxicology and the Environmental Mutagen Society. *Environmental and Molecular Mutagenesis* 14 (Suppl. 16): 1–6.

Weier, Anita. 2006. Mercury bigger risk for poor, minorities tend to eat more fish from Madison lakes. *Capital Times,* August 9.

Weijer, C. 1999. Protecting communities in research: Philosophical and pragmatic challenges. *Cambridge Quarterly of Healthcare Ethics* 8:501–513.

Weintraub, Max, and Linda S. Birnbaum. 2008. Catfish consumption as a contributor to elevated PCB levels in a non-Hispanic black subpopulation. *Environmental Research* 107 (3): 412–417.

Wenger, Etienne. 1998. *Communities of Practice: Learning, Meaning, and Identity*. Cambridge: Cambridge University Press.

Whaley, Rick, and Walter Bresette. 1994. *Walleye Warriors: An Effective Alliance against Racism and for the Earth*. Philadelphia: New Society Publishers.

Whiteside, Kerry H. 2006. *Precautionary Politics: Principle and Practice in Confronting Environmental Risk*. Cambridge, MA: MIT Press.

Williams, P. R. 2004. Health risk communication using comparative risk analyses. *Journal of Exposure Analysis and Environmental Epidemiology* 14 (7): 498–515.

Willis, Paul. 1977. *Learning to Labor: How Working Class Kids Get Working Class Jobs*. New York: Columbia University Press.

Wilson, Diane. 2005. *An Unreasonable Woman: A True Story of Shrimpers, Politicos, Polluters and the Fight for Seadrift Texas*. White River Junction, VT: Chelsea Green Publishing.

Winner, Langdon. 1986. *The Whale and the Reactor: A Search for Limits in an Age of High Technology*. Chicago: University of Chicago Press.

Wisconsin Department of Natural Resources. 2001. *Lower Rock River Water Quality Management Plan: Yahara River/Lake Monona Watershed*. Madison: Wisconsin Department of Natural Resources.

Wofford, Pamela, Randy Segawa, and Jay Schreider. 2009. Pesticide air monitoring in Parlier, CA. California Department of Pesticide Regulation. http://www.cdpr.ca.gov/docs/envjust/pilot_proj/parlier_final.pdf.

Wolfley, Jeanette. 1998. Ecological risk assessment and management: Their failure to value indigenous traditional ecological knowledge and protect tribal homelands. *Culture and Research Journal* 22 (2): 151–169.

Woodhouse, Edward J., and Dean Nieusma. 2001. Democratic expertise: Integrating knowledge, power, and participation. In Matthijs Hisschemoller, Rob Hoppe, William N. Dunn, and Jerry R. Ravetz, eds., *Knowledge, Power, and Participation in Environmental Policy Analysis*, 73–96. New Brunswick, NJ: Transaction.

World Wildlife Federation–UK. 2004. Bad blood? A survey of chemicals in the blood of European ministers. London. http://assets.panda.org/downloads/badbloodoctober2004.pdf.

Wynne, Brian. 1996a. May the sheep safely graze? A reflexive view of the expert-lay knowledge divide. In Scott Lash, Bronislaw Szerszynski, and Brian Wynne, eds., *Risk, Environment and Modernity: Towards a New Ecology*, 44–83. London: Sage.

Wynne, Brian. 1996b. Misunderstood misunderstandings: Social identities and public uptake of science. In A. Irwin and B. Wynne, eds., *Misunderstanding Science? The Public Reconstruction of Science and Technology*, 19–46. Cambridge: Cambridge University Press.

Wynne, Brian. 2001. Public lack of confidence in science? Have we understood its causes correctly? Paper presented at the 3rd Congress of the European Society for Agricultural and Food Ethics.

Yearley, Steven. 2005. *Making Sense of Science: Understanding the Social Study of Science*. London: Sage.

Zwick, David, and Marcy Benstock, eds. 1971. *Water Wasteland: Ralph Nader's Study Group Report on Water Pollution*. New York: Grossman.

About the Authors

Rebecca Gasior Altman is a Lecturer in the Community Health Program at Tufts University. She received her PhD in sociology from Brown University in 2008.

Julia Green Brody is Executive Director of the Silent Spring Institute and Adjunct Assistant Professor at the Brown University School of Medicine. She conducts research on breast cancer and the environment and on community-based research and public engagement in science. Her work has been funded by the National Science Foundation and the National Institutes of Health, among other agencies, and published in such journals as *Environmental Health Perspectives* and *Cancer*.

Phil Brown is a Professor of Sociology and Environmental Studies at Brown University. His most recent book is *Toxic Exposures: Contested Illnesses and the Environmental Health Movement* (Columbia University Press, 2007). His edited book (with Rachel Morello-Frosch and Stephen Zavestoski), *Contested Illnesses: Citizens, Science, and Health Social Movements*, will appear next year from the University of California Press. He codirects the Contested Illnesses Research Group. He directs the Community Outreach Core of Brown's Superfund Research Program, and the Community Outreach and Translation Core of Brown's Children's Environmental Health Center.

Benjamin Cohen is a Visiting Assistant Professor of Science, Technology, and Society at the University of Virginia. He works at the intersection of agroenvironmental history, the history of science, and STS. His book, *Notes from the Ground: Science, Soil, and Society in the American Countryside* (Yale University Press, 2009), examines the cultural conditions from which science and agriculture came together in the nineteenth century. Other work has appeared in *Environmental History, Historically Speaking, Agriculture and Human Values, Organization and Environment*, and *McSweeney's*.

Jason Delborne is an Assistant Professor of Liberal Arts and International Studies at the Colorado School of Mines. He completed his doctoral work in Environmental Science, Policy, and Management at the University of California, Berkeley. His research focuses on highly politicized scientific controversies, examining the practice of scientific dissent and efforts to democratize technoscientific governance. Delborne coedited *Controversies in Science and Technology: From*

Evolution to Energy (with Mary Ann Liebert, 2010) and has published other recent work in the social studies of science and public understanding of science.

Kim Fortun is an Associate Professor of Science and Technology Studies at Rensselaer Polytechnic Institute. Her research brings together STS and environmental justice by examining the dynamics of the environmental field in different historical and geographic contexts. She is the author of *Advocacy after Bhopal: Environmentalism, Disaster, New Global Orders* (University of Chicago Press, 2001), which was awarded the 2003 biannual Sharon Stephens Prize by the American Ethnological Society. Her recent work has appeared in the *American Anthropologist, Science as Culture, Ethics, Place and Environment, Osiris,* and *Design Issues.*

Scott Frickel is an Associate Professor of Sociology at Washington State University, where he studies science, environment, and social movements. He is the author of *Chemical Consequences: Environmental Mutagens, Scientist Activism, and the Rise of Genetic Toxicology* (Rutgers University Press, 2004) and coeditor of *The New Political Sociology of Science: Institutions, Networks, and Power* (University of Wisconsin Press, 2006).

Wyatt Galusky is an Associate Professor of Humanities and co-coordinator of the Science, Technology, and Society program at Morrisville State College. His research focuses on the mediating power of technology in terms of how humans see the natural world and themselves. His current projects involve online environmental activism, food technologies (particularly domesticated animals), and alternative energy systems.

Karen Hoffman is an Assistant Professor in the Department of Sociology and Anthropology at the University of Puerto Rico, Rio Piedras. Her work is focused on questions of science, technology, democracy, and social reproduction and change in the area of environmental policymaking and advocacy.

Jaclyn R. Johnson is an associate at Quarles & Brady LLP in Phoenix, Arizona, where she practices Indian Law and Real Estate Law. She received her BA from Dartmouth College, majoring in Native American Studies modified with Environmental Studies. She received her JD from the University of Michigan Law School. Johnson is a descendant of the Confederated Salish and Kootenai Tribes of the Flathead Nation.

Raoul S. Liévanos is a PhD student in the Department of Sociology at the University of California, Davis, who focuses on the social and spatial distribution of environmental hazards and the institutionalization of movements for environmental justice. Support for his research was provided by the Ford Foundation and the University of California Toxic Substances Research and Teaching Program through the Atmospheric Aerosols and Health Lead Campus Program.

Jonathan K. London is an Assistant Professor in the Department of Human and Community Development and the Director of the Center for Regional Change at the University of California, Davis. His research is on rural community development and environmental justice.

Rachel Morello-Frosch, PhD, MPH, holds a joint appointment as an Associate Professor at the University of California, Berkeley, in the Department of

Environmental Science, Policy, and Management and the School of Public Health. Her research examines race and class determinants of environmental health among diverse communities in the United States. One focus of her current work has been community-based health research, children's environmental health, the relationship between segregation and environmental health inequalities, and the intersection between economic restructuring and community environmental health.

Dean Nieusma is an Assistant Professor of Science and Technology Studies at Rensselaer Polytechnic Institute. His research focuses on social studies of design, development, and appropriate technology; renewable energy technology; and the relationship between expertise and democratic practices.

Gwen Ottinger is an Assistant Professor in the Interdisciplinary Arts and Sciences Program at the University of Washington, Bothell. Her work examines the construction of expertise in interactions between technical professionals and environmental justice activists around refinery fencelines and air-quality monitoring instruments.

Carla Pérez is Program Coordinator at Movement Generation Justice and Ecology Project, where she coordinates the community eco-justice training program, manages the Trainers Network, and leads the organization's Resiliency and Permaculture work. She has worked on issues of environmental justice and sustainable agriculture with community groups from Yucatán, Mexico, to San Francisco Bay Area environmental justice communities such as Richmond and Bay View Hunters Point. Pérez received her B.S. in Conservation and Resource Studies from the University of California, Berkeley.

Jim Powell is a cofounder of the Madison Environmental Justice Organization, Clean Air Madison, Dane County United, Grassroots Leadership College, Dane County Timebank, and several land conservation friends groups. He has organized successful community-led campaigns to improve pollution monitoring, food security, living wage, affordable housing, neighborhood safety, and community development, taught comparative religions at a community college, led many leadership development trainings, managed bookstores, run conference centers, and operated a ski chalet.

Maria Powell is a cofounder and Executive Director of the Madison Environmental Justice Organization and Research Director for the Nanotechnology Citizen Engagement Organization. She has a PhD from the University of Wisconsin, Madison, in Environmental Studies and Journalism/Mass Communication.

Darren Ranco, a member of the Penobscot Indian Nation, is an Associate Professor of Anthropology and Coordinator of Native American Research at the University of Maine. He completed his PhD in Social Anthropology from Harvard University (2000) on Penobscot critiques of EPA risk methodologies. His research focuses on the ways indigenous communities in the United States resist environmental destruction by using local knowledge to protect cultural resources, and how state knowledge systems, rooted in colonial contexts, continue to expose indigenous peoples to an inordinate amount of environmental risk. His recent publications have appeared in *Antipode*, *Society & Natural Resources*, and *Wicazo Sa Review*.

Ruthann Rudel   is Director of Research at the Silent Spring Institute, where she studies environmental pollutants, particularly endocrine-disrupting compounds, and women's health. She holds an adjunct appointment as a research associate in the Brown University Department of Pathology and Laboratory Medicine, and is active in the area of regulatory toxicology.

Julie Sze   is an Associate Professor of American Studies and Director of the John Muir Institute of the Environment: Environmental Justice Project at the University of California, Davis. Working at the intersection of American studies, environmental, urban, and ethnic studies, Sze's research focuses on race, class, gender, and the environment; the environmental justice movement; urban environmentalism; and environmental health.

Ami Zota   is a postdoctoral research fellow at University of California, San Francisco where she uses her expertise in epidemiology, exposure assessment, and environmental justice to investigate the cumulative impacts of environmental and social factors on reproductive health.

Index

Peggy F. Barlett and Geoffrey W. Chase, eds., *Sustainability on Campus: Stories and Strategies for Change*

Steve Lerner, *Diamond: A Struggle for Environmental Justice in Louisiana's Chemical Corridor*

Jason Corburn, *Street Science: Community Knowledge and Environmental Health Justice*

Peggy F. Barlett, ed., *Urban Place: Reconnecting with the Natural World*

David Naguib Pellow and Robert J. Brulle, eds., *Power, Justice, and the Environment: A Critical Appraisal of the Environmental Justice Movement*

Eran Ben-Joseph, *The Code of the City: Standards and the Hidden Language of Place Making*

Nancy J. Myers and Carolyn Raffensperger, eds., *Precautionary Tools for Reshaping Environmental Policy*

Kelly Sims Gallagher, *China Shifts Gears: Automakers, Oil, Pollution, and Development*

Kerry H. Whiteside, *Precautionary Politics: Principle and Practice in Confronting Environmental Risk*

Ronald Sandler and Phaedra C. Pezzullo, eds., *Environmental Justice and Environmentalism: The Social Justice Challenge to the Environmental Movement*

Julie Sze, *Noxious New York: The Racial Politics of Urban Health and Environmental Justice*

Robert D. Bullard, ed., *Growing Smarter: Achieving Livable Communities, Environmental Justice, and Regional Equity*

Ann Rappaport and Sarah Hammond Creighton, *Degrees That Matter: Climate Change and the University*

Michael Egan, *Barry Commoner and the Science of Survival: The Remaking of American Environmentalism*

David J. Hess, *Alternative Pathways in Science and Industry: Activism, Innovation, and the Environment in an Era of Globalization*

Peter F. Cannavò, *The Working Landscape: Founding, Preservation, and the Politics of Place*

Paul Stanton Kibel, ed., *Rivertown: Rethinking Urban Rivers*